CHINA: RAILWAY PATTERNS AND NATIONAL GOALS

by

LEUNG Chi-Keung
University of Hong Kong

THE UNIVERSITY OF CHICAGO
DEPARTMENT OF GEOGRAPHY
RESEARCH PAPER NO. 195

1980

Published 1980 by the Department of Geography
The University of Chicago, Chicago, Illinois

Library of Congress Cataloging in Publication Data

Leung, Chi-Keung.

China, railway patterns and national goals.

(Research paper—The University of Chicago, Department of Geography; no. 195)

Bibliography: p. 215

1. Railroads—China—History. 2. Railroads and state—China—History. I. Title. II. Series: Chicago. University. Dept. of Geography. Research paper; no. 195.

H31.C514 no. 195 [HE3288] 910s [385'.0951] 80-17030

ISBN 0-89065-102-7

Research Papers are available from:

The University of Chicago
Department of Geography
5828 S. University Avenue
Chicago, Illinois 60637
Price: $8.00 list; $6.00 series subscription.

To my wife,

Lai Ping

ACKNOWLEDGMENTS

The conception of this study dates back nearly a decade. It was Autumn 1970 in Chicago, when the so-called Chinese model of development began to be vaguely conceived by the outside world, and when the desire therefore to trace over space the logic of development in China appeared most challenging academically. Although the thoughts were soon transformed into a crude program of investigation in early 1971, heavy commitments during the years after my return to the Department of Extra-Mural Studies of the University of Hong Kong in 1971 to my move to the Department of Geography and Geology in 1973 prevented any sustained attempts at more than data collection, while ill-health constantly interrupted the analytical work, which was not started until Summer 1975. Fortunately, a year of sabbatical leave, 1976-1977, in London rendered a slow recovery possible. What is gratifying is that despite the intervening years the conception and organization of the study appear to stand, though some of its freshness might have been lost in the long delay.

Throughout these years, my debt and gratitude to Norton Ginsburg is immeasurable, not only for his generous assistance which brought me to Chicago, but also for his great confidence, understanding and patience in my work, his unfailing encouragement and support, and his numerous suggestions and criticisms on the first draft. I am also most grateful to Bert F. Hoselitz, Center of Economic Development and Cultural Change, who jointly financed my Fellowship at the University of Chicago; to Chauncy D. Harris and William D. Pattison for their careful review of the study; and to Tang Tsou for some early inspiration. While I will never forget Dawn and Lea whose help in Chicago tremendously smoothened our settling in, I am always sad in my affectionate memory of Dawn since her death in 1972.

I should like to record my thanks to the Center of Asian Studies, the University of Hong Kong, and to the Association of Commonwealth Universities. The former provided me with two generous grants, one for the purchase of research materials in 1971 and another for the employment of a research assistant for three months in the Summer of 1975. The latter awarded me a Commonwealth Academic Staff Fellowship, 1976-77, which enabled me to spend some time at the Public Record Office, London, while being attached to the

University of London. I have certainly benefitted from criticisms and suggestions from faculties, colleagues, fellow students, friends and others at seminars and discussions on the subject matter on the campuses of the University of Chicago, the University of London, Keele University and the University of Hong Kong.

Nau Kam Nam, my research assistant, reconstructed railway networks of various periods, compiled appendix B, and did most of the topological calculations for table 26. With the exceptions of figures 1, 8, 26, and 27 which have been drawn by Albert F. de Souza, Cartographic Studio, SOAS, University of London, all illustrations are drawn by H. K. Kwan, T. B. Wong, and Martin Chiu, Cartographic Unit, while photographic assistance has been provided by Y. S. Cheung and K. B. Chueng, Geological Laboratory, both of the Department of Geography and Geology, University of Hong Kong. Advice and assistance on computer programming have been received from N. Ng, and M. S. Luk, Computer Center, and the cover has been designed by Dimon Liu, School of Architecture. The entire draft, which at various stages comprised of heaps of papers, tables, and hardly legible scribbles, has been efficiently and neatly typed by Betty Lai. For the assistance and contributions of them all, I wish to record my sincere thanks. I must also thank the Wardens and other Quaker friends at the Friends International Center, London, for their warmth toward a convalescent who was drawn inward by his work and the desire to conserve his limited energy.

This study, however, would never have been possible without the tolerance of my old mother, and the support and sacrifices of my wife, Lai Ping, whose courage and devotion in adversity I have come to treasure. I hope the completion of this piece of work will represent a token of my sincere efforts to make up for fractions of all the deprivations that Lai Ping and our children, Eli, Eva, and E-feng, have endured over the years.

LEUNG Chi-Keung

Department of Geography & Geology
University of Hong Kong
April 1979

TABLE OF CONTENTS

ACKNOWLEDGMENTS v

LIST OF ILLUSTRATIONS xi

LIST OF TABLES. xiii

LIST OF PHOTOGRAPHS xv

Chapter

I. INTRODUCTION: TRANSPORTATION AND POLITICS. 1

Transportation Geography
Some Conceptual Models
Transportation and Development
Transportation and Political Geography

Iconography, Circulation, and the Space-Polity

The Chinese Case

PART ONE. RAILWAYS AND POLITICS IN PRE-COMMUNIST CHINA

II. RAILWAY PATTERNS AND POLITICS IN LATE CH'ING CHINA . . 17

Politics and Ideology in China's Initial Railway Development
Political and Ideological Background
Ideologies and the Location of Railways
The removal of the Wusung-Shanghai Railway
The construction and extension of the T'angshan-Hsukuochuang Railway
Conflicts over the Tientsin-Tungchou and Peking-Hankow Railways

International Politics and China's Railway Development
The "Scramble for Concessions"
The "Sphere of Influence" and the Doctrine of "Open Door"
The Russo-Japanese Struggle

Popular Movements and Nationalization of Railways
Private and Provincial Enterprises
Nationalization and the Fall of the Ch'ing Dynasty

Patterns and Functions of Railways in Imperial China
The Network Patterns and Characteristics
The Economics and Functions of Chinese Railways

Summary of Findings

III. RAILWAY PATTERNS, NATIONALISM, AND THE DEFEAT OF THE REPUBLIC REPUBLIC OF CHINA 51

Railway Development and National Aspirations
The Sun Yat-sen Doctrine and His Development Plans
Railway Strategies and Japanese Penetration

Railway Development and a Divided Space-Polity
Railway Strategies and Shrunken National Space
Japanese Railway Strategies in Manchuria

Railways and Wars
Railway Development and the Sino-Japanese War
Japanese Railway Strategies in Occupied China
Post-War Railway Plan and the Civil War

Patterns and Functions of Railways in the Republic of China
The Patterns and Functions of Nationalist Railways
The Patterns and Functions of a Colonial System

Summary of Findings

PART TWO. RAILWAY PATTERNS AND NATIONAL GOALS IN COMMUNIST CHINA

IV. RAILWAY TRANSPORTATION IN THE PEOPLE'S REPUBLIC OF CHINA. 93

Objectives and Data

Network Development
The Rehabilitation of the Disintegrated System, 1949-1952
The First Five-Year Plan, 1953-1957
The Second Five-Year Plan, 1958-1962
The Cultural Revolution and the Third Five-Year Plan, 1966-1970
The Fourth Five-Year Plan, 1971-1975

The Railway System and National Transportation
State Investment
Imports of Railway Equipment
Freight and Passenger Traffic
Railways and the Contraction of Space

Summary

V. RAILWAY PATTERNS AND NATIONAL GOALS: AN EXPLORATION . . 123

Communist Ideology and Transportation Development
Mao Tse-tung Thought and Developmental Contradictions

Spatial and Regional Contradictions
The political space
The physical space
Regional socio-economic differences
Transportation Planning

Railway Patterns and Chinese Space-Polity
Network Structure and Emphasis
Nodal Accessibility and Areal Organization
Railway Connectivity and Areal Hierarchy

In Search of Development Goals
Railway Input and Development Priorities
The data
The pre-Cultural Revolution goals
The post-Cultural Revolution goals
The railway network in 1975
Communist Development Strategy and Dialectics

Summary of Findings

PART THREE. CONCLUSION

VI. IDEOLOGY, TRANSPORTATION, AND SPACE-POLITY. 167

APPENDICES

A. FOREIGN RAILWAY CONCESSIONS IN CHINA, 1895-1911 & 1912-1914 173

B. RAILWAY TRANSPORTATION DATA 176

C. RAILWAY ACCESSIBILITY RANKINGS, 1949-1975 195

D. NOTES AND SOURCES FOR ILLUSTRATIONS 201

E. THE INDEPENDENT VARIABLES: SOURCES AND EXPLANATIONS . . 205

GLOSSARY. 209

SELECTED BIBLIOGRAPHY 215

LIST OF ILLUSTRATIONS

Figure		Page
1.	Location of the Tientsin-Tungchou and Peking-Hankow Railways.	27
2.	Railway Spheres of Influence of Foreign Powers	32
3.	Railways in China, 1906.	36
4.	Railways in China, 1911.	42
5.	Railways in China, 1930.	58
6.	Railways in China, 1937.	61
7.	Railways in China, 1945.	67
8.	China's Foreign Trade and Trade Deficit, 1864-1948	77
9.	Railways in China, 1949.	97
10.	Railways in China, 1952.	98
11.	Railways in China, 1957.	100
12.	Railways in China, 1963.	102
13.	Railways in China, 1970.	105
14.	Railways in China, 1975.	107
15.	Modern Transport Performance	112
16.	Passenger Train Frequency, 1956.	116
17.	Passenger Train Frequency, 1963.	117
18.	Railway Distance of Selected Pairs of Cities, 1956-1963.	119
19.	Railway Journey Time, 1956	120
20.	Railway Journey Time, 1963	121
21.	The Hostile Land of China.	129
22.	Distribution of Population and Operating Railway	130
23.	Railway Accessibility Rankings, 1949	138

Figure | Page

24. Railway Accessibility Rankings, 1963 139

25. Railway Accessibility Rankings, 1975 140

26. Distribution of Operating Railways in "Ecumenical" and "Extra-ecumenical" China 156

27. Distribution of Gross Value of Industrial Output in "Ecumenical" and "Extra-ecumenical" China 157

LIST OF TABLES

Table		Page
1.	Capitalization of Provincial Railway Companies	38
2.	Railway Construction by Private and Provincial Companies.	38
3.	Railway Construction in China, 1876-1911	40
4.	China's Foreign Trade by Principal ports, 1870-1948. . .	43
5.	Foreign Railway Loans to China, 1887-1911.	44
6.	Average Construction Costs of Railways in China.	46
7.	Foreign Control of Chinese Railways, 1894 and 1911 . . .	47
8.	Changes of Railway Kilometrage during the Sino-Japanese War (Outside Manchuria)	68
9.	Railway Construction in China, 1912-1948	73
10.	China's Railway Debts before and after World War I . . .	75
11.	Proportion of Railway Debt Payable to Net Income, 1912-1935.	76
12.	Operating Ratios of Chinese, Manchurian, and French Railways in China, 1908-1940	79
13.	Operating Conditions of Chinese Railways, 1917-1935. . .	80
14.	Military Transportation during the Sino-Japanese War . .	81
15.	Ownership and Control of Railways in Manchuria, 1930-1931.	83
16.	Capital-Output and Operating Ratios of Major Manchurian Railways, 1903-40, Selected Years	84
17.	Directions of the Chinese Eastern Railway's Freight Traffic, 1908-24, Selected Years	85
18.	Import-Export Ratios of the Manchurian Railways, 1925-1929.	86
19.	South Manchurian Railway Company, Profit & Loss, 1931-1932.	87

Table

Table		Page
20.	Railways and People's Liberation Army's War Progress, 1946-1949	95
21.	Railway Construction, 1950-1962	103
22.	State Investment in Transport and Communications, 1953-1960	109
23.	China's Imports of Machinery and Transportation Equipment	110
24.	Railway Freight Transport Performance, 1949-1975	113
25.	Railway Passenger Transport Performance, 1949-1960	115
26.	Topological Indices of Railway Networks in Mainland China, 1949-1975	133
27.	Railway Connectivity of National and Provincial Level Capitals	143
28.	The Independent Variables	147
29.	Correlation between "Railways Operating 1963" and Selected Independent Variables	149
30.	Correlation between "Railways added 1949-1963" and Selected Independent Variables	150
31.	Correlation between "Shift-and-share Index 1963-1975" and Selected Independent Variables	151
32.	Correlation between "Railways added 1949-1970" and Selected Independent Variables	152
33.	Correlation between "Shift-and-share Index 1963-1975 and Selected Independent Variables	153
34.	Correlation between "Railways Operating 1975" and Selected Independent Variables	154

LIST OF PHOTOGRAPHS

Plate Page

1. The Paot'ou-Lanchou Railway passing through the Tengri Desert 159

2. Sections of the Paochi-Ch'engtu Railway in mountainous Shenshi Province. 159

3. The Yingt'an-Hsiamen Railway. 160

4. The Ch'engtu-K'unming Railway overcoming steep gradient and difficult terrain. 161

5. The laying of 1000 m long rails on the Peking-Tientsin Line 162

6. Modern transportation facilities at the service of China's minority nationalities, the Yuanmou station on the Ch'engtu-K'unming Railway. 162

7. Railway builders participate in conservation work along the Chiaotso-Chihcheng Railway 163

8. A train runs through the tracks of the Southern Hsinchiang Railway, while construction obviously continues 163

All photographs by courtesy of Ta Kung Pao, Hong Kong.

CHAPTER I

INTRODUCTION: TRANSPORTATION AND POLITICS

Transportation Geography

Transportation is a subject studied by many disciplines. Defined as a service or facility which creates time and place utility through the physical transfer of persons and goods from one location to another, whereas production creates form utility, transportation interests economists mainly in its "intermediate" role linking production with consumption, or as an input of production.[1] As an important part of economic activity, the journey-to-work has been analyzed by sociologists - and some geographers - as a form of social interaction to assess its effects on the daily commuting public.[2] To the transportation engineers and the planners, traffic, whether goods or passenger traffic, flows between towns and between town and countryside because there are complementary activities generating cross-movements, and it is therefore a function of human activities, economic or otherwise.[3]

---

[1]According to John B. Lansing, Transportation and Economic Policy (New York: Free Press, 1966), pp. xi-xv. Thus, economic historians study the causes and consequences of transportation development and economic policies in the past, e.g., Robert W. Fogel; economic analysts analyze transportation legislation and other government public finance policies; economic theorists are concerned with the development of relevant theoretical tools, e.g., microeconomic theory, theory of public enterprise economics, and theory of investment, including some location theorists who study transportation in relation to industrial location. Of course, economists are also interested in industrial organization analysis, in engineering economics, and in the application of econometric techniques of analysis to transportation.

[2]Kate K. Leipmann, The Journey to Work (London: n.p., 1945). The journey-to-work is also studied by geographers, in terms of the commuting range and regional accessibility, see for example, Robert E. Dickinson, "The Geography of Commuting: The Netherlands and Belgium," Geographical Review 47 (October 1957): 521-538, and "The Geography of Commuting in West Germany," Annals of the Association of American Geographers 49 (December 1959): 443-456; and Richard E. Lonsdale, "Two North Carolina Commuting Patterns," Economic Geography 42 (April 1966): 114-138.

[3]Colin Buchanan, Traffic in Towns (London: Britain Ministry of Transport, 1963), pp. 33-34.

In Geography, transportation has long been a subject of systematic study as part of economic geography. In the early classics of locational theory, transportation was examined as an important factor of location either in agriculture or in industry ever since von Thünen and Weber.[1] Subsequent works by Hoover, Lösch, and Isard follow the same line though they have been largely " . . . to improve the spatial and regional frameworks" of locational theory.[2] In 1954, Ullman wrote that "transportation is a measure of the relations between areas," and that its study is to provide a deeper insight into the meaning of areal differences.[3] Thus "transportation facilities are examined primarily as indicators of the degree of connection and as patterns of spatial interchange," and he accordingly propounds three conditions under which interaction develops, namely, complementarity, intervening opportunity, and transferability,[4] which are often regarded as basic organizing concepts in transportation geography.[5] Current studies, those

---

[1]Maurice Fulton and L. Clinton Hoch, "Transportation Factors Affecting Location Decisions," Economic Geography 35 (January 1959): 51-59; and Peter Haggett, Locational Analysis in Human Geography (London: Edward Arnold Publishers Ltd., 1965), p. 13.

[2]Peter Haggett, ibid. See also, Edgar M. Hoover, The Location of Economic Activity (New York: McGraw-Hill Book Company, Inc., 1948); August Lösch, The Economics of Location (New Haven: Yale University Press, 1959); and Walter Isard, Location and Space-Economy: A General Theory Relating to Industrial Location, Market Areas, Land Use, Trade and Urban Structure (Cambridge: M.I.T. Press, 1956).

[3]Edward L. Ullman and Harold M. Mayer, "Transportation Geography," in American Geography: Inventory and Prospect, eds. Preston E. James and Clarence F. Jones (Syracuse: Syracuse University Press, 1954), p. 331; and Michael E. Eliot Hurst, Transportation Geography, Comments and Readings (New York: McGraw-Hill Book Company, Inc., 1974), p. 2. According to Hurst, this idea is still retained today, although in a more refined way, in the geometric and absolute spatial concepts of Bunge, Nystuen, in Berry's use of factor analysis to relate spatial structure and commodity flows, and in Haggett's attempts to erect an "integrated regional system" around movement, networks, nodes, hierarchies, and surfaces.

[4]Edward L. Ullman, "The Role of Transportation and the Bases for Interaction," in Man's Role in Changing the Face of the Earth, ed. W. L. Thomas (Chicago: University of Chicago Press, 1956), pp. 862-880.

[5]For example, see Ronald Abler, John S. Adams, and Peter Gould, Spatial Organization (Englewood Cliffs: Prentice-Hall, 1971), pp. 193-235; and Michael E. Eliot Hurst, Transportation Geography, Comments and Readings.

concerned with the pattern and function of the transportation networks,[1] with the establishment of some correlation between network structure and regional or national characteristics,[2] and with simulating such networks through computers by mathematical methods,[3] generally make use of graph theory and treat transportation networks as systems of nodes, routes, and surfaces, "as if the reality of our everyday world were in a sociopolitical vacuum."[4]

Some Conceptual Models

Haggett has attempted to fuse the sequence of network development in developed areas with a theoretical Löschian landscape. Haggett's model begins with an initial dense network of intersecting pathways connecting village settlements. In the second stage, smaller centers are by-passed by the longer interaction-distance routes constructed to link up major centers; whereas in the third stage the interaction has been raised still further with a new set of optimum routes connecting only very few largest centers. In

---

[1]For example, W. L. Garrison and Duane F. Marble, The Structure of Transportation Networks, Technical Report 62-11, (U.S. Army Transportation Command, 1962); Karel J. Kansky, Structure of Transportation Networks: Relationship between Network Geometry and Regional Characteristics, Research Papers, no. 84 (Chicago: University of Chicago Department of Geography, 1963); Nils Petter Gleditsch, The Structure of the International Airline Network (Oslo: mimeographed, 1968); Peter Haggett and Robert Chorley, Network Analysis in Geography (New York: St. Martin's Press, 1969); and R.B. Potts and R.M. Oliver, Flows in Transportation Networks (New York: Academic Press, 1972).

[2]For example, Brian J.L. Berry, "An Inductive Approach to the Regionalization of Economic Development," in Essays in Geography and Economic Development, ed. Norton S. Ginsburg, Research Papers, no. 62, (Chicago: University of Chicago Department of Geography, 1960), reproduced with modifications as "Basic Patterns of Economic Development," an appendix to Norton Ginsburg, An Atlas of Economic Development (Chicago: University of Chicago Press, 1961); Karel J. Kansky, Structure of Transportation Networks; and John D. Nystuen and Michael F. Dacey, "A Graph Theory Interpretation of Nodal Regions," in Spatial Analysis, ed. Brian J.L. Berry and Duane F. Marble (New Jersey: Prentice-Hall, 1968).

[3]For example, Karel J. Kansky, Structure of Transportation Networks; Richard L. Morrill, Migration and the Growth of Urban Settlement, Lund Studies in Geography, no. 26 (Royal University of Lund: Sweden, 1965); John Kolars and Henry J. Malin, "Population and Accessibility: An Analysis of Turkish Roads," Geographical Review 60 (April 1970): 229-246; and Budd Hebert and E. Murphy, "Evolution of an Accessibility Surface: The Case of the U.S. Air Transport Network," Proceedings of the Association of American Geographers Vol. 3 (1971), pp. 75-79.

[4]Michael E. Eliot Hurst, Transportation Geography, Comments and Readings, p. 2.

other words, Haggett postulates a network development by route substitution between successively higher-order centers, along with the emergence of an urban hierarchy.[1]

Perhaps the most often quoted model in transportation geography is that first stated by Fisher in 1948, and more recently refined by Gould, et al., of an "ideal-typical sequence" for the evolution of transport lines and towns in colonial seaboard areas. In this model, the authors identify four distinct phases: Phase I consists of a scattering of small ports along the coast in pre-colonial times; Phase II the colonial penetration lines from sea coast to the interior; Phase III the proliferation of feeder lines from the inland nodes to further exploitation of resources in the hinterlands; and Phase IV the establishment of a primate port city, supported by "high-priority corridors" of concentrated traffic from the interior.[2] This model has since been widely applied to other developing countries,[3] though the exact meaning of its underlying theory has been discussed only recently and remains unclear.[4]

---

[1]Peter Haggett, Locational Analysis in Human Geography, pp. 82-83, based on W.L. Garrison, et al., Studies of Highway Development and Geographic Change (Seattle: University of Washington Press, 1959).

[2]C.A. Fisher, "The Railway Geography of British Malaya," The Scottish Geographic Magazine 64 (1948): 123-136; Peter R. Gould, The Development of the Transportation Pattern in Ghana, Northwestern University Studies in Geography, no. 5 (Evanston; Northwestern University Press, 1960); and Edward J. Taaffe, Richard L. Morrill and Peter Gould, "Transport Expansion in Underdeveloped Countries: A Comparative Analysis," Geographical Review 53 (October 1963): 503-559. Since the model is widely quoted, no attempt will be made to reproduce it here.

[3]For example, Ward's application of the model to Malaya, Stanley's to Liberia, and Briggs to East Africa. See Marion Ward, "Progress in Transport Geography," in Trends in Geography, ed. R.U. Cooke and J.H. Johnson (Oxford: Pergamon Press, 1969), pp. 164-173; W.R. Stanley, "Transport Expansion in Liberia," Geographical Review 60 (October 1970): 529-547; and J.A. Briggs, "The Development of the East African Railway Network - An Application of a Sequential Development Model," Swansea Geographer. 12 (1974): 47-49.

[4]For example, H.C. Brookfield in 1973 points out the parallel between this "ideal-typical sequence" model for the evolution of transport lines and towns and the "stages of growth' of Walter W. Rostow, while contrasting the independent and simultaneous development of a spatial framework known as the "center-periphery" model by J. Friedmann. Brookfield argues that they, in conformity with the diffusionist school of development theory, assume "a contagious diffusion process." See H.C. Brookfield, "On One Geography and a Third World," Transactions 58 (March 1973): 1-20.

Transportation and Development

In addition, a variety of other perspectives have developed on the role transportation plays in development.[1] The first concerns the question as to whether improved transportation is a prerequisite or a permissive condition for economic growth. In trying to establish the stimulus for the take-off stage of economic growth in the United States, Rostow identifies the railroad as the critical investment sector.[2] Contrary to this view, Cootner argues that railroad growth after 1830 followed rather than preceded the growth of other sectors of the economy,[3] and Fogel concludes that the railway, as opposed to road and inland waterways, was only marginally more economical and effective as a means of overland transportation, and was therefore not indispensable.[4] It should perhaps be noted that Fishlow, using a similar approach, arrives at a different answer - that the railroad was important to the economic transformation of the American economy, especially the agricultural sector.[5]

---

[1]For example, Wilfred Owen, Strategy for Mobility (Washington, D.C.: Brookings Institution Press, 1964); Holland Hunter, Soviet Transport Experience (Washington, D.C.: Brookings Institution Press, 1968); Patrick O'Sullivan, Transport Networks and the Irish Economy (London: London School of Economics and Political Science Press, 1969); Peter J. Rimmer, Transport in Thailand. Department of Human Geography, (Canberra: Australian National University Press, 1971); B.S. Hoyle, Transport and Development (New York: Barnes & Noble, 1973); and Thomas R. Leinbach, "Transportation and the Development of Mayala," Annals of the Association of American Geographers 65 (June 1975): 270-282.

[2]Walter W. Rostow, The Stages of Economic Growth (Cambridge: Cambridge University Press, 1964), p. 24. Similarly, Hunter suggests that the effect of transportation innovation has been to widen markets and to permit economies of large-scale production, and that thus there is a causal relationship between transportation and economic development; whereas Owen regards transportation as the key to national development on the same ground. See Holland Hunter, "Transport in Soviet and Chinese Development," Economic Development and Cultural Change. 14 (1965): 71-72; and Wilfred Owen, Strategy for Mobility, p. 2.

[3]P. Cootner, "The Role of the Railroads in the U.S. Economic Growth," Journal of Economic History 23 (December 1963): 477-521.

[4]Robert W. Fogel, Railroads and American Economic Growth (Baltimore: Johns Hopkins Press, 1964).

[5]See Albert Fishlow, American Railroads and the Transformation of the Ante-Bellum Economy (Cambridge: Harvard University Press, 1965).

The second theme concerns the spatial structure of development resulting from transportation improvement.[1] Transportation innovation has long been perceived to stimulate the diffusion of economic development in space, particularly from urban to rural areas, or from center to periphery, and thus to promote spatial equilibrium.[2]

Today, although some studies may still choose to regard transportation as the lead factor,[3] the general consensus seems to be that transportation is only a concomitant of, not a precondition for, economic growth, in capitalist as well as in socialist economy.[4] In terms of the spatial consequences,

> One conspicuous result of effects of improved transportation on spatial interaction and regionalization is the increasing dominance of the largest cities . . . The village has lost functions to the town, the town to the city, and the metropolis has usurped many of the functions of the satellite cities surrounding it.[5]

More evidence is now available to show that, "transportation

---

[1]The slight overlap with the Peter Gould, et al. model, should be noted. However, the emphasis here is the dynamic process resulting from transportation innovation rather than the evolution of the transport network pattern per se. For examples of such studies, see footnote 1, page 5 above. See also footnote 4, page 4 above for a brief discussion on the parallel of the underlying assumption and therefore the area of the overlap.

[2]For example, Berry stresses the relationship between the development of a central-place system and a state of entropy in a socio-economic system, achieved in the steady-state of a stochastic process, whereas Friedmann believes that the spatial objective of economic development is the progressive replacement of a center-periphery structure with a single system of cities extending throughout the economic space. See Brian J.L. Berry, "Cities and Systems within Systems of Cities," in Regional Development and Planning, eds. W. Alonso and J. Friedmann, (Cambridge: M.I.T. Press, 1964), pp. 116-137; and J. Friedmann, "An Approach to Policies Planning for Spatial Development," School of Architecture and Urban Planning, (Los Angeles: University of California at Los Angeles, July, 1974).

[3]For example, Peter J. Rimmer, Transport in Thailand, p. 7.

[4]Robert W. Fogel, Railroads and American Economic Growth; A.M. O'Connor, Railways and Development in Uganda, A Study in Economic Geography (Nairobi: Oxford University Press, 1965); Arthur Rosenbaum, "Railway Enterprise and Economic Development, the Case of the Imperial Railways of North China, 1900-1911," Modern China 2 (1976), pp. 227-272; and for an example in socialist economy, see Holland Hunter, Soviet Transport Experience, p. 123.

[5]Edward J. Taaffe, "The Transportation Network and the Changing American Landscape," in Problems and Trends in American Geography, ed. Saul B. Cohen (New York: Basic Books, 1967), p. 23.

investment strengthens the center-periphery structure of the economy rather than generating a movement toward spatial integration.[1]

Transportation and Political Geography

It is obvious from the preceding discussion that transportation has been treated, in geography as in other disciplines, essentially as an economic activity, be it a locational factor, a measure of areal differences, an indicator of technological and therefore economic level, a lead or permissive factor in the process of economic development. In short, it is firmly entrenched in the chain concept of production-transportation-consumption of the economists. Thus Wolfe in 1963 complained that "when transportation is the main subject (of discussion), it is usually thought of as an economic activity."[2] Paradoxically, it was Cooley, a pioneer in modern transportation research, who in 1894 argued against such a bias, pointing out that, "There can be no adequate theory of transportation which has regard only to some one aspect of its social function, as the economic aspect."[3] He believed that in order to develop an adequate theory of transportation, "one must examine severally its relations to various social institutions, . . . military, political, economic, and ideal . . . ,"[4] yet his emphasis was clearly on political institutions. To Wolfe, political events and political institutions always involve men interacting over distance, thus there is a distinct and intimate relationship between politics and transportation, whether transportation is used as an offensive instrument by the Romans and Incas, or as a means of colonial exploitation.[5]

---

[1]Howard L. Gauthier, "Geography, Transportation and Regional Development," in Transport and Development, ed. B.S. Hoyle (New York: Barnes & Noble, 1973), pp. 19-31. For evidence, see William V. Ackerman, "Development Strategy for Cuyo, Argentina," Annals of the Association of American Geographers 65 (March 1975): 36-47; and P.C. Forer, "Relative Space and Regional Imbalance: Domestic Airlines in New Zealand's Geometrodynamics," Proceedings of the International Geographical Union Regional Conference, New Zealand (1974), pp. 53-62.

[2]Roy I. Wolfe, Transportation and Politics (Princeton: Van Nostrand, 1963), p. 3.

[3]Charles H. Cooley, The Theory of Transportation (Baltimore: American Economic Association, 1894), p. 44.

[4]Ibid.

[5]Roy I. Wolfe, Transportation and Politics, p. 7.

Such a relationship is more obvious when transportation is discussed at the national rather than the corporate level. For in war or in peace, "few things are more important to the nation than sound transportation."[1] Whereas Goodrich argues that government plays a more important role than private enterprises in promoting growth through transportation, especially for developing countries;[2] Haefele, in examining transportation planning and national goals in the developed world, sees that, "every transportation choice is intertwined with choices in other sectors and that they all depend, ultimately, on overall national decisions," and that, "the problem of national goal setting . . . are also problems of establishing values.[3] Moreover, Hurst in his study of transportation and the Canadian landscape, points out that, "at the basis of the economic geography of that landscape lies capitalism, its values and its identities." To him, the reality is an aspect of the political-economic system sometimes called "transportation politics".[4]

Iconography, Circulation, and the Space-Polity

The attention to transportation politics has had a long tradition in Political Geography, although most texts on the subject pay only lip-service to it.[5]

In 1890, Mahan, advocating a thesis of the peripheral (or maritime) versus the central (or continental), recognized the importance of circulation[6] to political power.[7] More explicitly,

---

[1]Karl M. Ruppenthal, ed., Challenge to Transportation (Stanford: Stanford University Press, 1961), p. v.

[2]Carter Goodrich, Government Promotion of American Canals and Railroads, 1800-1890 (New York: Columbia University Press, 1960).

[3]Edward T. Haefele, ed., Transportation Planning and National Goals (Washington, D.C.: Brookings Institution Press, 1969), p. 177-178.

[4]Michael E. Eliot Hurst, "The Railway Epoch and the Canadian Landscape: Case Studies in Transportation and the Societal Milieu," manuscript of a chapter for T. O'Riordan, ed., American Landscape (to be published).

[5]For example, see A.E. Moodie, Geography Behind Politics (London: Hutchinson, 1949); Lucile Carlson, Geography and World Politics (Englewood Cliffs: Prentice-Hall, 1958); and Paul Buckholts, Political Geography (New York: Ronald Press, 1966).

[6]As conceived by the French geographer, Vidal de la Blache, "circulation" comprises transportation and communication, which shade into each other but are not identical.

[7]A.T. Mahan, The Problem of Asia (Boston: Little, Brown and Company, 1900).

Mackinder places mobility at the center of international politics, arguing that modern history exhibits a "counteraction" with maritime peoples developing superior mobility over a different kind of surface (sea versus land) and controlling margins of Euro-Asia, where no navigable rivers penetrate, and where the steppe peoples are inconquerable. However, he already saw the potential possibility of the railway in again reversing the situation of mobility differentials between the peripheral and continental states to the advantage of the latter.[1] With development in air transportation technology, de Seversky in 1950 held that the land and sea forces of the world had been subordinated to the greater mobility of the air.[2] These large historical and geographical generalizations as Mackinder calls them[3] not only show that political and strategic power depends on a superior circulation system, but also demonstrate that transportation innovation necessarily alters the political organization of area on a global scale.

At the nation-state level, that there is on the other hand a close relationship between transportation and the state-idea also is widely accepted. According to Hartshorne, who emphasizes a functional rather than a historical or evolutionary approach to Political Geography, especially in studies of political areas, the state is the product of two opposing forces, the centrifugal and the centripetal, and what is primarily responsible for the integration and formation of political communities is the existence of what he calls the "state-idea" (or "raison d'etre").[4] Similarly, Gottmann postulates that political areas of the world are differentiated on two elements - "circulation" and "iconography," which in terms of a given state are also fundamental to its coherence.[5]

---

[1]H.J. Mackinder, "The Geographical Pivot of History," Geographical Journal 23 (1904): 421-444.

[2]Alexander P. de Seversky, Air Power: Key to Survival (New York: Simon & Schuster, 1950).

[3]H.J. Mackinder, "The Geographical Pivot of History."

[4]Richard Hartshorne, "The Functional Approach in Political Geography," Annals of the Association of American Geographers 40 (June 1950): 95-130.

[5]Jean Gottmann, "The Political Partitioning of the World: An Attempt at an Analysis," World Politics 4 (July 1952): 512-519; and "Geography and International Relations," World Politics. 3 (March 1951): 153-173.

Circulation includes the movement of men, ideas, goods, armies, and capital, whereas iconography, analogous to Hartshorne's state-idea, consists of all those common memories, symbols, and experiences cherished by a particular people.[1] For the first time and against the geographers' obsession with physical space, Gottmann stresses that the most stubborn boundaries of the political world consist not of physical barriers but barriers of the spirit, thus the main political partitions in space are those that have resulted in the minds of people.

Although both Hartshorne and Gottmann singled out only iconography (state-idea) and circulation (integrating and disintegrating processes), their relevance to political areas is clearly taken for granted. However, it was only when Jones developed his unified field theory in 1954 that iconography, circulation, and the space-polity were conceptually linked.[2] Specifically, the model consists of five interlocking components:

Political Idea - Decision - Movement - Field - Political Area

Here "political idea" can be roughly defined as "state-idea" or "iconography"; "movement" as "circulation"; and "political area" is exactly Hartshorne's "functional area" or Gottmann's "political partition", or simply the "space-polity".[3] More important, perhaps, is Jones' emphasis that the five components of the model are like a chain of lakes interconnected at one level and that, instead of being a linear one, the model is more like a matrix, so that

---

[1]Soja interpretes Gottmann's concepts of circulation and iconography in the following manner, " . . . circulation, the dynamic movement of goods, people, and ideas which permits space to be organized but concurrently makes for fluidity and change; . . . iconography, the array of symbolic phenomena which resists the change brought about by movement to favor a certain established order or pattern." See Edward W. Soja, The Political Organization of Space (Washington, D.C.: Association of American Geographers, 1971), p. 34.

[2]Stephen B. Jones, "A Unified Field Theory of Political Geography," Annals of the Association of American Geographers. 44 (June 1954): 111-123.

[3]Following Joseph Whitney, China, Area, Administration, and Nation Building, Research Papers, no. 123 (Chicago: University of Chicago Department of Geography, 1970), p. 7, "space-polity" is used to denote a system of relationships in a political area.

whatever enters one component would automatically spread to all the others. Circulation, being the formal and dynamic link between space-polity and iconography, is therefore part and parcel of all political considerations and activities.

So far, the usefulness of Jones' model as a whole remains untested, since studies carried out before and after its publication tend to handle only segments of the model: some examining the relationship between political idea and communication, especially in studies of integration and disintegration;[1] some examining that between circulation and political areas;[2] and few examining that between political idea and political areas.[3] In most cases, transportation is only a peripheral interest, and the spatial dimension is totally absent in non-geographic studies.

This study, in which transportation is the center of interest, will attempt to make use of the whole of the unified field theory and will therefore probe in both directions toward iconography and space-polity. Yet, although transportation is the center of interest, it is by no means a conventional transportation study.

The Chinese Case

The case of Chinese railway system affords an excellent example for such study primarily because its development has been shaped by

---

[1]For example, Richard Hartshorne, "The Functional Approach in Political Geography"; Jean Gottmann, "The Political Partitioning of the World"; Karl W. Deutsch, Nationalism and Social Communication (Cambridge: M.I.T. Press, 1966); Amitai Etzioni, ed., Political Unification: A Comparative Study of Leaders and Forces (New York: Holt, Rinehart, & Winston, 1965); Alan P.L. Liu, Communications and National Integration in Communist China (New York: John Wiley and Sons, 1964); Lucian Pye, Communication and Political Development (Princeton, New Jersey: Princeton University Press, 1963).

[2]For example, A.T. Mahan, The Problem of Asia; H.J. Mackinder, "The Geographical Pivot of History"; A.P. de Seversky, Air Power: Key to Survival; Edward W. Soja, "Communications and Territorial Integration in East Africa," The East Lakes Geographer 4 (December 1968): 39-57; David Hilling, "Politics and Transportation: the Problems of West Africa's Landlocked States," in Essays in Political Geography, ed., C.A. Fisher (London: Methuen, 1968), pp. 253-269.

[3]For example, Theodore Herman, "Group Values toward the National Space: The Case of China," Geographical Review 40 (April 1959): 164-182; Joseph B.R. Whitney, China, Area, Administration, and Nation Building; and Saul Cohen, Jerusalem: Bridging the Four Walls (New York: Herzl Press, 1977).

three political regimes with apparently different ideologies and aspirations, and has been subjected to foreign political influence until recently. Furthermore, of all modern facilities for inland transportation in China,[1] the railway system is clearly the most important in terms of both freight and passenger movements in spite of its small mileage.[2] Thus some of the more important questions that this study seeks to answer are: What have been the characteristics of, and differences in, railway network development and railway transportation in Imperial, Nationalist, and Communist China? What have been the socio-political forces working for or against railway construction during these periods? Have there been distinct sets of political aspirations or national goals, and have these been reflected in the political organization of area under different regimes?

Unfortunately, Imperial China has never been known for her statistical collection, much less for those relating to railway operation. Continuous disruptions during the Republican period prevented any efforts at nation-wide systematic collection of railway statistics.[3] Communist China, on the other hand, seldom publishes any economic data of significance, let alone specific information on railway operations, especially after the 1960's. Generally, therefore, there is a dearth of continuous nationwide data on freight and passenger movements, revenues and expenditures, equipment, etc. Where such information is available, it is more likely to be non-directional, regional or local, or for only a few selected years. To overcome some of these problems, the present study, drawing on more readily available materials such as those

---

[1]Facilities for inland transportation in China are extremely modest, whether in terms of railway, road, or waterway mileage, or in terms of density per unit of area or mileage per capita. See Norton Ginsburg, An Atlas of Economic Development, pp. 60-77.

[2]The railway system from 1949-1958 consistently carried some 80% of the total ton-kilometrage of freight movement, and more than 70% of the total person-kilometrage of passenger movement in China, though the rate of increase of highway passengers has been greater. See Wei-ta ti shih-nien [The great ten years] (Peking: 1959), pp. 131 and 133.

[3]Yet the Republican period was one with most abundant, though fragmentary and scattered, data. Recently the Public Record Office in London declassified Foreign Office files on China up to the mid-1940's, rendering more information available. Unfortunately, statistical tables attached to reports in these files have been removed.

relating to routing, foreign loans, treaties and agreements concerning railway construction, railway plans and designs, etc., focuses on network development and the resultant spatial patterns from the appearance of the first railway in China in 1876. Reference to operational, including traffic, characteristics will be made only where possible and relevant.

PART ONE

RAILWAYS AND POLITICS IN PRE-COMMUNIST CHINA

CHAPTER II

RAILWAY PATTERNS AND POLITICS IN LATE CH'ING CHINA

Politics and Ideology in China's Initial Railway Development

Political and Ideological Background

From the beginning the creation of railway lines in China has been a political and ideological issue. The long decline of the Ch'ing Dynasty since the late Eighteenth Century reached its point of no return after the first Opium War (1840-42) and the signing of the Nanking Treaty,[1] by which China not only had to establish free trade Treaty ports, to surrender customs control, and to grant extraterritoriality, but China was rendered a semi-feudal and semi-colonial country. Thereafter, China launched herself on the strenuous course of modernization, whereby oppression and anti-oppression campaigns[2] followed one another and, in terms of internal politics, the means of modernization and national defence were central to all political and ideological struggles. The railroad was obviously one of such means.

The earliest public advocate for railway construction in China was the Kan Wang (Shield Prince) of the Taiping Heavenly Kingdom, who in 1859 stated,

> To promote transportation: it should emphasize convenience and speed, . . . e.g., trains in foreign countries, which travel 7-8,000 li in one day . . . First the twenty-one provinces should be connected . . . so that it serves as the trunk road, for connection ensures the health of the state . . .[3]

---

[1]Ting Ming-nan, et al., Ti-kuo chu-i chin hua shih [History of imperialist invasions of China] Vol. 1, (Peking: Jen-min chu-pan-she, 1961), pp. 9-11.

[2]To name only a few chronologically, the Two Opium Wars (1840-42, 1856-60), the Taiping Revolution (1851-64), the Sino-French War 1883-85), the Sino-Japanese War (1894-95), the Reform Movement (1898), the Yi Ho Tuan Movement (1900-01), the Russo-Japanese War (1904-05), etc.

[3]Hung Jen-kan, Tzu cheng hsin-pien [New guide to government], as quoted in Li Kuo-ch'i, Chung-kuo tsao-ch'i ti t'ieh-lu ching-ying [Early railway development in China], (Taipei: Academia Sinica, 1961), p. 10.

After the Taiping Revolution (1851-64), an important part of the reformist movement to learn from the West concerned the role of the railway. The proponents[1] advocated that since the railway was an effective weapon of the colonial powers, China could not defend herself without understanding and possessing the same weapon. Moreover, it would fulfill the need for national integration and therefore internal security, and for modernization. The opponents,[2] however, argued that once constructed, the railway would effectively penetrate the interior, deprive the populace of its livelihood, and ultimately render China exposed and defenseless.

Meanwhile, converging on the China scene in the treaty ports, the colonial sea powers rapidly found, in continental China, that not only the effectiveness of the steamship was seriously restricted by the limited navigable waterways, but also that a superior means of movement was required to counteract each other's encroachment upon the vast and relatively impenetrable hinterland.[3] The railroad was no doubt ideal for the purpose. Thus as early as 1858, only two years after her occupation of lower Burma, Britain was entertaining plans for railway penetration from Rangoon into Central China simply because,[4]

> . . . its completion would permanently preserve to . . . [British] manufacturers and merchants that equality of position in the China trade, of which both the Americans and Russians aim so strenuously trying to deprive them, by their Pacific Ocean approaches toward Eastern Asia.[5]

Nevertheless, the introduction of the 'fire car' (or train) into China, in which process the colonial powers took forceful

---

[1]Including first Wang Tao, Yung Hung, Liu Ming-ch'uan, Ting Jih-ch'ang, and later Hsueh Fu-ch'eng, Ma Chien-chung, Li Hung-chang, Chang Chih-tung, Tso Tsung-tang, T'ang T'ing-shu, and many others.

[2]Including Chang Chia-hsiang, Liu K'un-i, Liu Hsi-hung, etc. See Cheng Lin, The Chinese Railways (Shanghai: China United Press, 1935), p. 12.

[3]See also George Moseley, "The Frontier Regions in China's Recent International Politics," in Modern China's Search for a Political Form, ed. J. Gray, (London: Oxford University Press, 1969), pp. 299-329.

[4]Ch'en Hui, Chung-kuo t'ieh-lu wen-ti [China's railway problems], (Peking: San-luan Books, 1955), p. 21.

[5]Captain Sprye's Rangoon Scheme, 1858-1866. (F.O. 17/470-471).

initiative while China responded doggedly,[1] brought in a whole range of problems resultant from the meeting of two conflicting sets of ideologies and aspirations, each rooted deeply in their respective cultural systems.

Ideologies and the Location of Railways

The Removal of the Wusung-Shanghai Railway. From 1858 onward, the British made numerous efforts to build railroads in China,[2] including a comprehensive plan by MacDonald Stephenson who envisaged an east-west oriented system linking China with Burma and India, but totally ignored the serious inadequacies of north-south communications in China.[3] In 1863, a total of 27 firms of British,[4] American, and French origin in Shanghai jointly petitioned for permission to construct a line between Suchou and Shanghai.[5] The petition immediately aroused suspicion and apprehension among Chinese officials, and Li Hung-chang, then Viceroy of Chiangsu, was directed

---

[1]This idea is extended from Tang Tsou's theme of "American initiative and Chinese response" in Sino-American relationship. See also Tang Tsou, American Failure in China, 1941-50, 2 vols. (Chicago: University of Chicago Press, 1967).

[2]In 1862, a formal application was initiated by W. F. Mayers, a British resident in Kuangtung for the construction of a line from Kuangtung to Chianghsi province. In the winter of the same year, John Mcleary Brown, et al., assigned by the British embassador to investigate the coal reserves in the Peking vicinity, recommended the construction of a railway line. In July 1864, a British merchant constructed for demonstration a toy line about one third of a mile long outside the Hsuanwu Gate, Peking, which was soon removed. See Li Kuo-ch'i, Chung-kuo tsao-ch'i ti t'ieh-lu ching-ying, p. 12.

[3]MacDonald Stephenson, "Report upon the Feasibility and Most Effectual Means of Introducing Railway Communication into the Empire of China", (F.O. 233/78, 1864).

[4]The 27 firms involved were Messrs. W. R. Anderson & Co.; Birley, Worthington & Co.; Dent & Co.; Gibb Livingston & Co.; Gilman & Co.; Holiday, Wise & Co.; Jarvin Thornburn & Co.; Lindsay & Co.; Russell & Co.; E. M. Smith; Thorne Brothers & Co.; Watson & Co.; Geo. Barnet & Co.; Bower, Hanbury & Co.; Fletcher & Co.; Hogg Brothers & Co.; Augustine Heard & Co; Jardine Matheson & Co.; Johnson & Co.; Olyphant & Co.; Reiss & Co.; Remi, Schmidt & Co.; David Sasson, Sons & Co.; Smith Kennedy & Co.; Turner & Co.; Wheelock & Co.; and Alfred Wilkinson & Co. See Fu Yu-cheng, Chung-kuo chin-t'ai tieh-lu shih tzu-liao, 1863-1911 [Modern railway history materials of China, 1863-1911] (Peking: China Books, 1963), pp. 1308-09.

[5]According to Percy H. B. Kent, Railway Enterprise in China (London: Edward Arnold, 1907), pp. 3-4, the petition was made on July 20, 1863 to Li Hung-chang. However, according to Fu Yu-ch'eng, Chung-kuo chun-t'ai tieh-lu shih tzu-liao, 1863-1911, p. 3, quoting from archives of the Tsungli Yamen, Li was informed by the kuan-tao of Shanghai only between November and December of that year.

by the Tsungli Yamen to absolutely reject any such applications from foreigners.[1] In his reply, Li Hung-chang clearly stated that the construction of railways was deemed to be beneficial to China only when undertaken by the Chinese themselves; that serious objection existed against the expropriation of land for railroad construction; and that there was strong dislike of the employment of foreigners in the interior.[2] Privately, Li was suspicious of the motives behind the collaboration of the three colonial powers,[3] and he felt that the petition was designed to further the political rather than commercial objectives of its promoters.[4]

In spite of this, only two years later British merchants in Shanghai formed the Woosung (Wusung) Railway Company,[5] whose failure to make any progress for many years led to its absorption by the Woosung (Wusung) Road Company, apparently a subsidiary of the Jardine, Matheson and Company, which decided in 1872 to take the matter into its own hands. Under the disguise of building a road, the Company started in December 1874 route preparation between Shanghai and Wusung, while rails and other equipment did not arrive until early 1876.[6] As soon as the first segment of the railway line appeared, the Taotai of Shanghai, Feng Chun-kuang, called on the

---

[1]See Li Kuo-ch'i, Chung-kuo tsao-ch'i ti t'ieh-lu ching-ying, p. 12.

[2]Chang, Kia-ngau, China's Struggle for Railway Development (New York: John Day, 1943).

[3]Li Hung-Chang's letter to Tsungli Yamen of February, 1864, see Fu Yu-ch'eng, Chung-kuo chin-t'ai tieh-lu shih tzu-liao, 1863-1911. pp. 4-5, quoting from archives; and Lu Kuo-ch'i, Chung-kuo tsao-ch'i ti t'ieh-lu ching-ying, p. 12.

[4]Cheng Lin, The Chinese Railways (Shanghai: China United Press, 1935), p. 1.

[5]The Company appointed Henry Robinson as Chief Engineer and applied for permission to construct a railway line from Shanghai to Wusung. The application was absolutely rejected on the grounds of "seven undesirables" by the then Acting Taotai, Ying Pao-shih. See Li Kuo-ch'i, Chung-kuo tsao-ch'i ti t'ieh-lu ching-ying, p. 14.

[6]Hsu Ti-ch'en, "The Wusung Railway Negotiation", quoted in Fu Yu-Ch'eng, Chung-kuo chin-t'ai tieh-lu shih tzu-liao, 1863-1911. p. 38. See also T. Dennett, Americans in East Asia (New York: Barnes & Noble, 1922), pp. 595-596.

British consul, Walter H. Medhurst, on February 22 for an explanation.[1] Medhurst first argued that since land had been leased, he did not see it right to interfere as to its specific use. Medhurst further argued that since permission had been obtained for the construction of a road and since the importation of railway equipment had been duty-exempted, the Chinese government had no right to object. Feng stated that the permission was never granted for a railway, and that since the equipment was declared as necessary for the road, it was duty-exempted for the same reason. Henceforth, Feng rejected further requests for duty-exemption and refused to endorse all outstanding title deeds for land acquisition by the Company. Subsequently, many notes were exchanged between the two countries. When an impasse was reached, construction work was speeded up[2] to accomplish a _fait accompli_[3] in spite of Chinese opposition. On June 30, 1876, the completed line[4] was fully open to traffic, and nothing seemed possible to reverse the situation until a Chinese pedestrian was knocked down and instantly killed by the train on August 3. On September 13, Li Hung-chang and Thomas Wade met at Yent'ai to begin negotiation for a settlement. Although Britain was fully aware that the line could not be defended legally,[5] China, instead of forcefully closing the illegal line, offered to purchase it for Chinese disposal. Accordingly, the British

---

[1]Fu Yu-ch'eng, _Chung-kuo chin t'ai tieh-lu shih tzu-liao, 1863-1911_. pp. 41-42.

[2]The fact that the railway could have been completed seems to show that the British must have by-passed all official channels in their subsequent dealings.

[3]North China Herald Office, _A Retrospect of Political and Commercial Affairs in China, 1868-1872_, quoted in Fu Yu-ch'eng, _Chung-kuo chin t'ai tieh-lu shih tzu-liao, 1863-1911_. p. 36.

[4]The line measured 9¼ miles, had a gauge of 2.6 feet, and rail weight of 26 pounds per yard. It was therefore a light railway. When the line was open, the Company had two passenger engines, one small engine, ten passenger cars, and twelve four-wheeled freight cars. See commercial reports from the British Consul, Shanghai, 1876, p. 20-21 as quoted in Fu Yu-ch'eng, _Chung-kuo chin t'ai tieh-lu shih tzu-liao, 1863-1911_. pp. 37-38.

[5]Wade privately communicated Jardine, Matheson and Company that since the railway had been built without official approval, it could not be defended by the British Government. See Edward LeFevour, _Western Enterprise in Late Ching China, A Selective Survey of Jardine, Matheson and Company's Operations, 1842-1895_ (Cambridge: Harvard University Press, 1968), p. 109.

Government pressed for continued management of the line by the British company as a precondition of redemption,[1] while the American Ambassador tried to justify the sincerity of its promoters and urged its retention.[2] Eventually, an agreement was signed at Nanking on October 24, 1876, by which the Chinese government was to redeem the line by three instalments within one year, while the Company was allowed to continue the service until the last instalment was paid. When the redemption money of 285,000 taels was paid up in October 1877, China promptly removed the line and shipped all its equipment to Taiwan,[3] being convinced that,

> . . . the removal of the Wusung Railway is being carried out solely in consequence of the political necessity of the act. That to allow it to remain where it is would utterly stultify the action of the authorities and afford the strongest encouragement to similar invasions of Chinese territory and of her independence as a nation.[4]

Thus, the first construction in China, initiated by foreign powers and built 49 years after the world's first railway, the Stockton-Darlington Line, was destroyed. However, it would be wrong to attribute its destruction to negative Chinese response alone, for the British merchant subterfuge, encouraged by colonial interests, proved to be too much on a matter of strategic concern to China.

The construction and extension of the T'angshan-Hsukuochuang Railway. The removal of the Wusung-Shanghai railway highlighted only the ideological conflicts between the Chinese and the colonialists over the ownership, timing, and location of the line, rather than Chinese disinterest in railway development. It has been

---

[1]Li Kuo-ch'i, Chung-kuo tsao-ch'i t'ieh-lu ching-ying, p. 41.

[2]American embassador's memorandum to China, October 17, 1877, see Fu Yu-ch'eng, Chung-kuo chin-t'ai tieh-lu shih tzu-liao, 1863-1911. p. 58, quoting from Chu Shih-chia ed., Shih-chiu Shih-chi mei-kuo chin hua tang-an shih-liao hsuan-ch'i [Selection of historical documents of American invasion of China in 19th century] (1959), pp. 413-414.

[3]According to Fu Yu-ch'eng, Chung-kuo chin t'ai tieh-lu shih tzu-liao, 1863-1911, p. 58, footnote 1, the line was eventually re-shipped to North China for the construction of the Kaiping railway. Apparently, the British still managed to remove and ship some of the dismantled equipment back to Ipswich. See R. S. Lewis, Eighty Years of Enterprise, 1869-1949, Being the Ultimate Story of the Waterside Works of Ransomes and Papier Ltd. of Ipswich (Ipswich: n.p.), pp. 19-24.

[4]T. Dennett, Americans in East Asia, pp. 596-597, footnote.

pointed out that since the early 1860's, the Chinese had been seriously considering the manner in which railways were to be introduced into China. But the political and economic climate was hardly encouraging, for,

> . . . after decades of defeat and insult, national wealth has been exhausted, national welfare destroyed . . . (foreign) clocks and toys find their way into every household; foreign fibres and clothes into the remotest and poorest areas; and the trend in Chiangsu and Chechiang is to use foreign currencies, even at incredible prices, unashamedly . . .[1]

Knowing the situation, Li Hung-Chang[2] saw the preconditions for railway development from the very beginning,

> . . . the trial construction of railway . . . if suggested for China Proper, will most certainly meet strong criticism . . . Railway requires too large a capital layout to rely on the initiative of the public . . . (and) it would be absolutely hopeless if such a gigantic venture had to depend on imported materials . . .[3]

Li therefore believed that a railway program must commence with the mining of coal and iron ore. Thus in 1876 he started the Kaiping coal mine with T'ang T'ing-shu, and a year later submitted through T'ang a petition for constructing a light railway to transport coal from Kaiping to Ch'ienhok'ou.[4] Having failed to obtain Imperial sanction, T'ang resubmitted in 1880 a plan to improve the transportation of coal, on which the Chinese navy became dependent, including the digging of a canal between Lut'ai and Hsukuochuang and a light railway from Hsukuochuang to T'angshan, the coal mine, with trains drawn by horses. Construction started on June 9, 1881, and was

---

[1]Letter from Kuo Sung-t'ao to Li Hung-Chang, May 23, 1877, as quoted in Li Kuo-ch'i, Chung-kuo tsao-ch'i ti t'ieh-lu ching-ying, p. 46.

[2]Li Hung-chang had as early as 1874, when China was preparing to build her modern naval defence force and the establishment of the Board of Admiralty, immediately memorialised for railway construction. See Ch'en Hui, Chung-kuo t'ieh-lu wen-ti, p. 22.

[3]Li Hung-chang, Li Wen-chung-kung ch'uan-chi [Complete works of Li Hung-chang], Vol. 17, (n.p., n.d.) p. 13.

[4]Chiao-tung shih lu-cheng pien [History of communications: road and railway administration] Vol. 1, (Nanking, n.p., 1935), pp. 11-12.

completed by the end of the same year.[1]

Meanwhile, the Chinese advocates became bolder and more positive in their visions for railway development. For example, Liu Ming-chuan, in his December 1880 memorial, visualised the most comprehensive railway plan for China, with trunk lines from Peking reaching Ch'ingchiang in the east, Hankow in the central, Shenyang in the north, and Kansu in the west, and stated plainly its political significance:

> . . . there is no lack of armies, nor supplies. . . . but the territory is divided into eighteen provinces. If railways were built . . . the eighteen provinces will be united . . . so that the Imperial Court will centralize all military power . . . and will not be weakened by territorial officialdom.[2]

However, the chief impulses which led to official acceptance of the railway as a modern system of transportation were the imminent threat of the Russian encroachment on the Northeast and China's defeat in the Sino-French War of 1884, which brought home to the Imperial Government the railway's value as a means of military mobilization and national defence. Foreign interests, which never completely disappeared, soon became overwhelming,[3] and the Chinese advocates had to find a way to avoid foreign entanglement as well as to silence opposition. When the T'angshan-Hsukuochuang railway reached Yenchuang in 1886, and in response to Russian threat, Prince Ch'un explicitly instructed Li Hung-chang to push its extension further to the Tientsin-Taku area on the following grounds,

> . . . railways though beneficial, have yet to be freed from gossip and criticism . . . . Could it not be memorialised by the Board of Admiralty . . . that the section from Yenchuang to Taku . . . will be

---

[1]Ibid. For a discussion of the dates of the commencement and completion of construction, see Li Kuo-chi, Chung-kuo tsao-ch'i ti t'ieh-lu ching-ying, pp. 51-53. Also, according to Li, although it was intended to be a narrow-gauge line, standard gauge was eventually adopted. C. W. Kinder, the resident mining engineer, too, built with scrap materials a small locomotive capable of hauling 100 tons, which was used to pull trains between T'angshan and Hsukuochuang ever since.

[2]Liu Ming-chuan, Liu Chuang-shu kung tso yi [Memorials by Liu Ming-chuan] 2: 1-3. Strongly supporting Liu's memorial, Li Hung-chang advocated the well-known nine advantages of railway development more or less on the same line. See Li Hung-chang, Li Wen-chung-kung ch'uan-chi 39, pp. 20-26. Liu, in contrast to Li, suggested the building of the Peking-Chingchiang line in the first phase, and the use of foreign loans.

[3]For a description of the numerous efforts made by Britain, France, and Germany during this period, see Li Kuo-ch'i, Chung-kuo tsao-ch'i ti t'ieh-lu ching-ying, pp. 63-64.

> constructed by the Board for the transportation of marines and ammunition. Say that it is only an experiment, in future . . . (it might) gradually expand.[1]

Li added in his justification that the extension, lying tens of li inland would be strategically safe.[2] The line was eventually allowed to extend on its northeastern end beyond Shanhaikuan, and later became part of the Peking-Shenyang Railway. Its motive was so obvious that foreigners then commented that,

> That this plan was undoubtedly based on political interests can be seen from the fact that it was focused on the Government's obligation and responsibility to Manchu in the northeast and north . . . when implemented, (the plan) in fact would provide only the northern part of the Empire the benefit of railways.[3]

It was no accident therefore that the location of the first successful Chinese railway construction was away from the most densely populated areas of China and along a narrow coastal corridor joining Chihli and the Manchu homeland.

Conflicts over the Tientsin-Tungchou and Peking-Hankow Railways. As soon as the Yenchuang to Tientsin line was completed in October 1888, Li Hung-chang again suggested its extension into the Capital area terminating at T'ungchou.[4] Li's purpose was to improve communication between Peking and Tientsin, an important part of his railway plan for national defence.[5] In spite of the Imperial sanction, the gathering force of the opposition obliged the Imperial Court to review its decision. Ironically, opposition was likewise based on defence considerations: that Tungchou being the gateway of the Capital, the construction of the line would lead to exposure; that railways should be trial built in peripheral rather than in the core area, and so on.[6] When the generals and viceroys in the coastal provinces were thus consulted, Chang Chih-tung, then Viceroy of Kuangtung, made the alternative proposal of constructing the Peking to Hankow line. Chang criticised the policy of building railways along the coast or border areas, upon where the colonial powers were

---

[1]Fu Yu-ch'eng, Chung-kuo chin-t'ai tieh-lu shih tzu-liao, 1863-1911, p. 128.

[2]Chiao-tung shih lu-cheng pien, pp. 42-44.

[3]North China Herald, 1884, p. 700.

[4]Li Hung-chang, Li Wen-chung-kung ch'uan-chi 3, p. 8.

[5]Li Kuo-ch'i, Chung-kuo tsao-ch'i ti tieh-lu ching-ying, p.75.

[6]Ibid., pp. 76-77

encroaching, for it would be too long a defensive line. Rather he emphasized that the Peking-Hankow line, being in the interior, would unite and control many provinces, and would provide the necessary military mobility (Figure 1).[1] No doubt, locating a line in densely populated areas to ensure its feasibility and profitability was a new demension in China'a railway planning. Unfortunately, when Chang's plan was approved in lieu of the Tientsin-T'ungchou extension, China was forced by growing Russian and Japanese interests in Manchuria and Korea, the latter still a protectorate of China, to re-divert her extremely limited financial resources for the building of the Peking-Shenyang line[2] for more immediate strategic purpose, and the construction of the Peking-Hankow line was not undertaken until a later date.

However, long before the Peking-Shenyang line reached the designed eastern terminus for frontier defence, the Sino-Japanese War (1894-95) broke out, which ushered in a new era of colonial struggle in China.

International Politics and China's Railway Development

As has been pointed out, another important condition in China's railway development was her international relations. Article VII of the Sino-American Treaty of July 28, 1868, part of which reads, " . . . in regard to the construction of railroads, (etc.) . . . China reserves . . . the right to decide the time and manner and circumstances of introducing such improvements within the dominions,"[3] probably reflected some degree of respect for national sovereign rights at that time. Although under great pressure of foreign interests and agitation that continued after the removal of the Wusung-Shanghai line, China was able to plan and develop her railway network fairly independently.

---

[1]Chang Chih-tung, Chang wen-hsiang-kung ts'ou-kao [Memorials of Chang Chih-tung] 17 (1920): 3-9. It should also be noted that this is the same as the south trunk line in Liu Ming-chuan's grand plan, and the argument for its construction was identical.

[2]Then more popularly known as the intra-mural and extra-mural lines, comprising the T'ungchou-Shanhaikuan and the Shanhaikuan-Shenyang railways.

[3]Additional Article to the Treaty of Commerce, June 18, 1858 signed on July 28, 1868, by William H. Seward, Anson Burlingame, Chih Kang and Sun Chia-ku, as quoted in Cheng Lin, The Chinese Railways, p. 19.

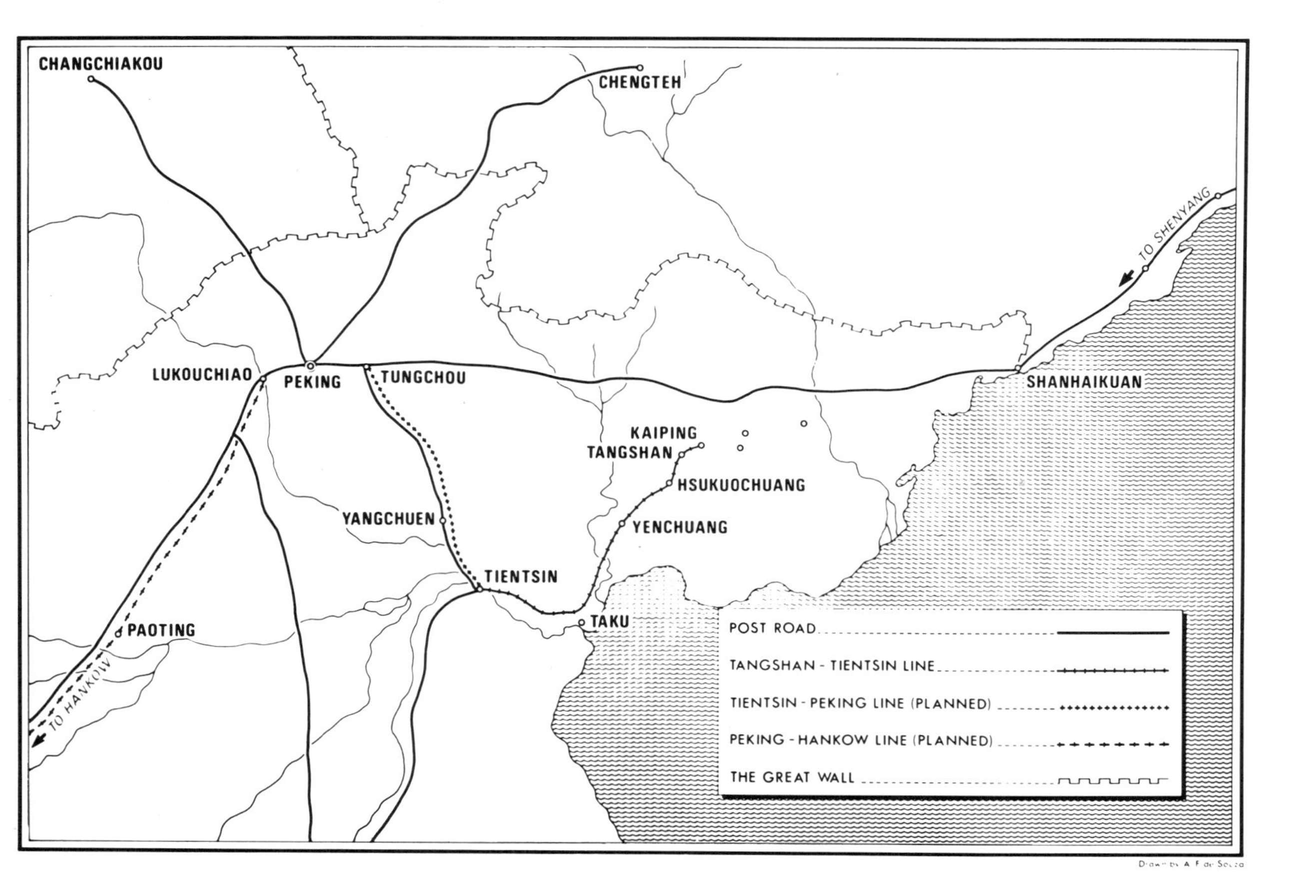

Fig. 1. Location of the Tientsin-Tungchou and Peking-Hankow Railways For sources, see appendix D.

This condition changed after the Sino-French War and the signing of the Treaty of Peace, Friendship, and Commerce of June 9, 1885,[1] which the French chose to interpret to include exclusive privileges in order to interfere with the construction of the Kuantung line in Manchuria from 1890 onward.[2]

The "Scramble for Concessions"

China's international status further deteriorated on signing the Treaty of Shimonoseki on April 18, 1895, after the Sino-Japanese War, which compelled China to recognise the independence of Korea, and to cede to Japan Port Arthur, Weihaiwei, and the islands of Taiwan and the Pescadores. A few months later, the *Dreibund*,[3] composed of Russia, Germany, and France, forced China to pay an increased indemnity of 230 million taels[4] to compensate Japan's retroceding Port Arthur and Weihaiwei, which eventually went to Russia and Britain respectively, while France obtained the right

---

[1]Article VII in French has been quoted in Cheng Lin, *The Chinese Railways*, p. 20, an English translation of which reads, "In order to develop in the most advantageous way the commercial and neighborly relations between France and China which the present treaty aims to re-establish, the government of the Republic will construct roads in Tonkin and will encourage the construction of railways in that area.

"When, for her part, China decides to construct railroads, it is understood that she will look to French industry for assistance and the government of the Republic will afford her every available facility for procuring necessary personnel in France. It is also understood that this clause cannot be considered as forming an exclusive privilege in France's favour." (the author wishes to acknowledge the kind assistance of Peter Collin, a former colleague at the University of Hong Kong, in the translation.)

[2]French interference in fact started earlier when the Peking-Hankow line was proposed in 1889. For a detailed discussion of the French interference in the construction of the Kuantung, or extra-mural, line in Manchuria see Li Kuo-ch'i, *Chung-kuo tsao-ch'i ti t'ieh-lu ching-ying*, pp. 85-98.

[3]In the Liaotung Convention, November 8, 1895, the *Dreibund* forced the Japanese government to retrocede Port Arthur and Weihaiwei in an attempt to exclude Japanese influence from North China.

[4]Knowing that China had difficulty paying the indemnity, M. de Witte, the Czar's Minister of Finance, devised a scheme of making the Russian government China's guarantor for the latter to float a loan of 400 million francs at 4 per cent interest, issued at 94 per cent, in Paris.

"to continue 'on Chinese territory either those railways already in existence or those projected in Annam' . . ."[1] A year later, China was lured into a joint venture, while Li Hung-chang[2] was in Moscow, to build the Chinese Eastern Railway on the ground of mutual defence,[3] which practically enabled Russia to send troops freely across Manchuria.

These events gave real impetus to the so-called "scramble for concessions." Thus, a year later, on the excuse of the murder of two German missionaries in Shantung and other incidents, Germany obtained the lease of Chiaochou and concessions to build railway lines from Chiaochou to Chinan, the provincial capital. On the ground that she was entitled to similar treatment, Russia leased Port Arthur and the Liaotung peninsula, and soon after edged the construction of the South Manchuria Railway linking the Chinese Eastern Railway at Harbin with Port Arthur. The _Dreibund_'s design on China was so subtle that even a loan which China had taken care to obtain in 1897 from Belgium for the construction of the Peking-Hankow line turned out to have been backed by Russia and France.[4]

Britain was so concerned about these developments that she mobilised her naval forces and demanded that China unconditionally accept the concessions of rights for five railway lines,[5] which were later regarded as "the fruit of the battle for concessions" for Britain. Japan too in 1900 unilaterally claimed exclusive rights for railway construction in Fuchien and adjacent areas.

---

[1]George E. Sokolsky, _The Story of the Chinese Eastern Railway_ (Shanghai: North China Daily News & Herald Ltd., 1929), p. 12. See also John V. A. MacMurray, ed., _Treaties and Agreements With and Concerning China, 1894-1919_ 1 (London: Oxford University Press, 1921), pp. 29-30. Moreover, according to Ch'en Hui, _Chung-kuo t'ieh-lu wen-ti_ p. 27, Britain quickly obtained the Burma-Yunnan railway concession by referring to Article IV of the Anglo-French Siam Convention of 1896 for mutual privileges in Yunnan and Ssuch'uan.

[2]In the Spring of 1896, when China was considering to send Wang Chih-ch'un as China's representative to the Coronation of Nicholas II, Count Cassini, the Russian Minister to Peking, insisted that Li Hung-chang, the negotiator for the Treaty of Shimonoseki, was most desirable.

[3]The Treaty of Alliance, May 1896, see John V. A. MacMurray, _Treaties and Agreements With and Concerning China, 1894-1919_, p. 81.

[4]Ch'en Hui, _Chung-kuo t'ieh-lu wen-ti_, p. 27.

[5]Ch'en Hui, _Chung-kuo t'ieh-lu wen-ti_, p. 28.

Appendix A is a summary of railway concessions the colonial powers exacted from China between 1895 to 1911. It will be seen that during the initial period of "scramble for concessions" from 1895 to 1905, Britain, France, Russia, and Germany played a leading role. Even though some of these rights were never exercised, the rapidity with which the colonial powers acquired them and the comprehensive coverage of Chinese territory testify not simply to the declining semi-colonial status of China, but also to the severity of the struggle amongst the powers themselves.

The "Sphere of Influence" and the Doctrine of "Open Door"

The "scramble for concessions" inevitably led to serious conflicts of interests among the colonial powers. The solution of conflicts among contemporary colonialists through the operation of the concept of "sphere of influence" begot conflicts at a later stage between the pioneers and the late-comers.

Britain and France: The earliest document pertaining to the concept of the sphere of influence was the Anglo-French Siam Convention of 1896, by which the two nations pledged each other to enjoy in common all privileges and advantages of any nature conceded to either nation within Yunnan and Ssuch'uan provinces.[1] That this agreement was hardly more than a *modus vivendi*,[2] and the objective behind far more than merely commercial, is clearly demonstrated by MacDonald's emphatic denunciation of French pretensions.[3]

Britain and Germany: To safeguard their mutual interests in the planned Tientsin-Pukou railway, which was to traverse Shantung province and the Yangtzu valley, and to prevent encroachment by Russia upon central China, the two countries agreed in 1898 to a demarcation of their sphere of influence over central and north

---

[1]John V. A. MacMurray ed., *Treaties and Agreements*, pp. 54-55, "Declaration with regard to the Kingdom of Siam, advantages in Yunnan and Ssuch'uan, etc., January 15, 1896," signed by Salisbury and Alph. de Courcel.

[2]Mongton Chih Hsu, "Railway Problems in China," (Ph.D. Dissertation, Colombia University, 1915), p. 45.

[3]*British Blue Book, Affairs of China*, no. 1, 1899, pp. 344-347, inclosure 2, MacDonald's despatch to Beresford.

China.[1] This arrangement created a German sphere of interest in Shantung and the Huang River valley which became a wedge between those of the British and the Russian.

Britain and Russia: Consequent to conflicts between Russia and Britain over the construction of the Shanhaikuan extension, the two countries agreed in 1899 that,

> Russia engages not to seek for herself or on behalf of Russian subjects other railway concessions in the Yangtzu Basin, and not to place obstacles either directly or indirectly in the way of railway enterprises in that region supported by the British government.
> (similar engagement, mutatis mutandis, by Britain with regard to the region north of the Great Wall.)[2]

However, after the Yi Ho T'uan Movement of 1900, the British extended their influence to Chihli, controlling trunk lines south of Shanhaikuan, while the Russians were confined to Manchuria north of this fort city.[3]

With the establishment of the "spheres of influence", China's territorial space was effectively divided and controlled by the Russians in Manchuria, the Germans in Shantung, the French in Yunnan and Kuanghsi, and the British in the Yangtzu valley (fig. 2). Moreover, a closed system was developed whereby all privileges were likely to be controlled or shared only by these powers. The monopolistic nature of railways contributed greatly to the situation.[4]

---

[1]This agreement, proposed by Herr M. A. von Hanseman of the German Syndicate, and accepted by the British and the Chinese Corporation and the Hong Kong and Shanghai Banking Corporation, was sanctioned by their respective governments in the "British and German Agreement held at New Court, St. Swithen's Lane, London, on the 1st and 2nd September, 1898," as quoted in Mongton Chih Hsu, "Railway Problems in China," p. 47 and footnoted.

[2]In the winter of 1898 when the Chinese Imperial Railway Administration concluded with the British and Chinese Corporation the Shanhaikuan Extension Railway Loan of ₤2,300,000, the Russian government raised serious objections to the Loan Contract while the British government of course upheld it. This was followed by a storm of violent protests and long diplomatic negotiations between London and St. Petersburg. On April 29, 1899, notes were finally exchanged and signed by the British Ambassador, Charles Scott, and Russian Minister for Foreign Affairs, Count Mourvaieff, at St. Petersburg. For the full texts of the agreement, see British Blue Book, p. 87.

[3]Percy H. Kent, Railway Enterprise in China, pp. 64-65.

[4]Ming-chien J. Pao, The Open Door Doctrine (New York: Macmillan Company, 1923), pp. 155-156.

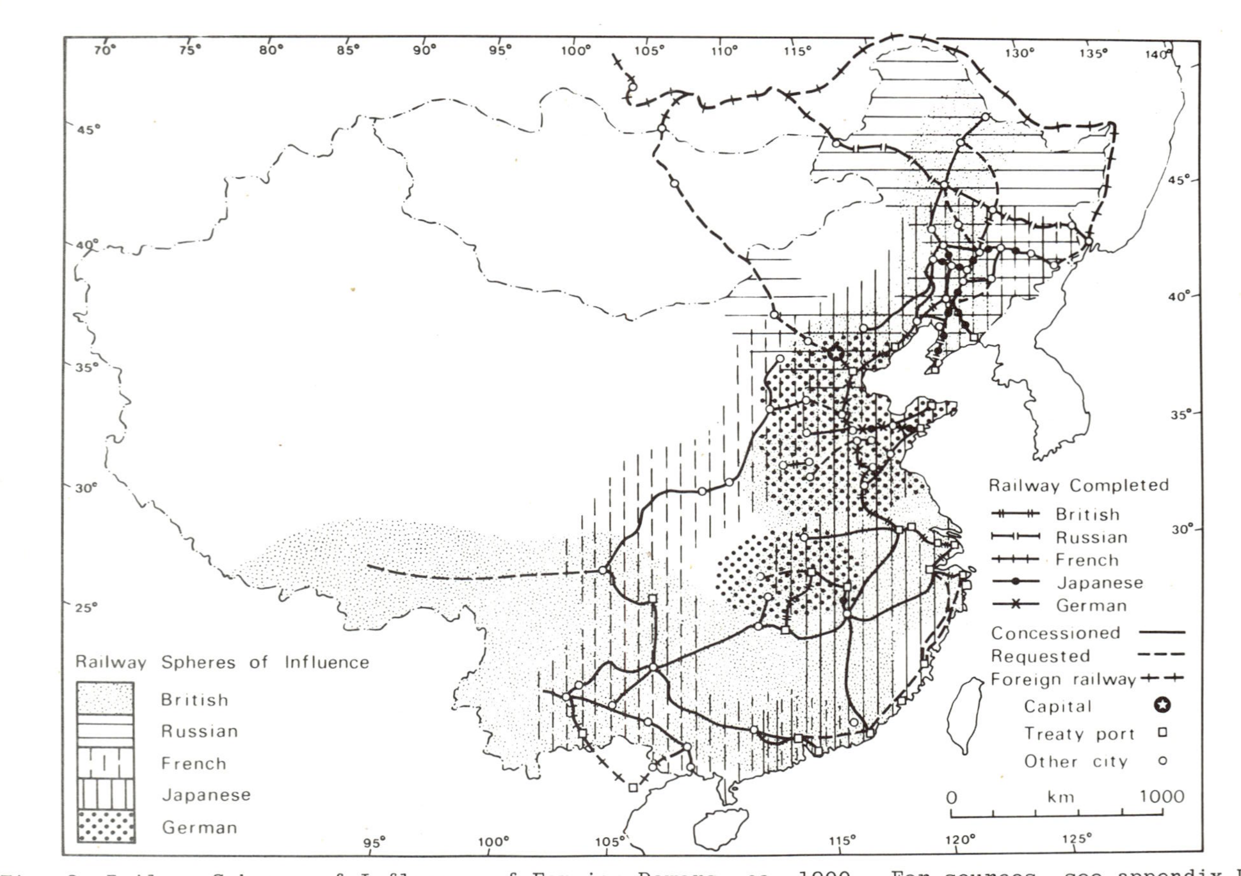

Fig. 2. Railway Spheres of Influence of Foreign Powers, ca. 1900   For sources, see appendix D.

This certainly did not please the United States, a latecomer to the drama of scramble in China. Hence, in response to the "unequal opportunities" created by the "sphere of influence", John Hay on September 6, 1899, re-emphasized the so-called Open Door Doctrine,[1]

> This government is animated by a sincere desire that the interests of our citizens may not be prejudiced through exclusive treatment by any of the controlling powers within their so-called 'sphere of interest' in China, and hopes also to retain there an open market for the commerce of the world, remove dangerous sources of international irritation . . .[2]

In spite of its avowed respect for vested interests, therefore, " . . . the Open Door Doctrine was enunciated to counteract . . . the spheres of influence, which menaced the reign of the equal opportunities of trade . . ."[3]

The course of events over the next few decades, in China Proper in general and in Manchuria in particular, showed that although the Doctrine essentially challenged only the monopolistic aspect of the "sphere of influence" rather than colonialism as such, it could not be accommodated by orthodox colonialism which was expansionary and exclusive in nature.

The Russo-Japanese Struggle

After the Yi Ho T'uan Movement, Russia seized all railways in Manchuria. By the time the Chinese Eastern and the South Manchurian railways were completed in 1903, Russia's "sphere of influence" was so firmly established in Manchuria that she was already looking

---

[1]The so-called "open-door policy" was proclaimed by Anson Burlingame as early as 1867. See Shutaro Tomimas, The Open Door Policy and Territorial Integrity of China (New York: Red House Press, 1918), pp. 42-43.

[2]U.S. Foreign Relations, 1899, pp. 131-133, John Hay to Choate, September 6, 1899.

[3]Ming-chien J. Pao, The Open Door Doctrine, p. 140. Akira Iriye, Across the Pacific (New York: Harcourt, Brace & World, 1967), pp. 80-81 and xi, argues that John Hay's policy had its domestic basis and suggests that Hay was on the defensive against the American expansionists of 1898.

toward north China for further expansion.[1] These developments were most unpalatable to Britain, which had vast political and commercial interests in north and central China;[2] to Japan, which was anxious to expand her colonial empire beyond Korea; and to the United States, champion of the open-door principle. With strong support from Britain and an obviously neutral United States, Japan was ready to seek revenge on the Dreibund for her humiliation at Shimonoseki. Contending that Russia was creating conditions prejudicial to Japan, destroying the principle of equal opportunity, and impairing the territorial integrity of China, Japan declared war on Russia on February 10, 1904. At the end of the war, the United States, trying to uphold the principle of equal opportunity, prevented victorious Japan from demanding an indemnity from Russia at Portsmouth, but instead agreed to transferring the Russian lease of the Liaotung peninsula and all the rights of the South Manchurian Railway from Ch'angch'un to Talien and Port Arthur to Japan. The principal clauses affecting the railway were:[3]

> Article V. The Imperial Russian Government transfer and assign to the Imperial Government of Japan . . . the lease of Port Arthur, Talienwan and adjacent territory and territorial waters and all rights, privileges and concessions connected with or forming part of such lease, and . . . also . . . all public works and properties in the territory . . .
> Article VI. The Imperial Russian Government engage to transfer and assign to the Imperial Government of Japan, without compensation . . . the railway between Ch'angsh'un (Kuanchengtzu) and Port Arthur and all its branches, together with all rights, priveleges and properties appertaining thereto in that region, as well as all coal mines in the said region belonging to or worked for the benefit of railway . . .

---

[1]George E. Sokolsky, The Story of the Chinese Eastern Railway, pp. 26-27, states that, "Witte once showed a map of China to Sir N. O'Conor on which he pointed to Manchuria, Chihli, Kansu, and Shenhsi as the probable areas which Russia would, in due course, absorb. It was apparently the Russian intention, in the extension of the Chinese Eastern Railway through Manchuria, to use it as a weapon . . . to penetrate China . . . to prevent any other Power from gaining a foothold . . . in any of the Chinese provinces bordering Russia".

[2]Ibid., pp. 27-28. The British policy over the question of Manchuria was best outlined by the speech of Michael Hicks-Beach, Chancellor of the Exchequer in 1898, " . . . We look upon it (China) as the most hopeful place of the future for the commerce of our country and the commerce of the world at large, and the Government was absolutely determined, at whatever cost even - . . . - if necessary at the cost of war, that the door should not be shut against us."

[3]John V. A. MacMurray, Treaties and Agreements, pp. 522-523, "Russia and Japan: Treaty of Peace, September 5, 1905".

China was forced to ratify the arrangement a few months later,[1] and to accept Japan's rights over the Antung-Shenyang railway which the latter had constructed during the Russo-Japanese War for military transportation without Chinese consent.[2] Two years later, Japan, probably driven by a desire to safequard whatever privileges she had obtained, entered into an agreement with Russia for the preservation of the status quo of the two countries in the area. Thereafter, Japanese interests grew not only in Manchuria[3] but also in other parts of China, as is evidenced by the increasing number of concessions she obtained (appendix A).

Popular Movements and Nationalization of Railways

Private and Provincial Enterprises

Toward the end of the Nineteenth Century, the populace of China had become alarmed by the avidity of the colonial powers, and disillusioned by the Ch'ing government.[4] By 1906, when the dust of the Russo-Japanese War was settled, almost the entire railway network of some 5,500 kilometers was under foreign control directly or indirectly (figure 3). Widespread dissatisfaction soon led to movements for the protection and redemption of railway rights, and for the development of provincial and private railways,[5] both in line with the reformist movement for "self-strengthening."

After the construction of the first private line, the Canton-

---

[1]Ibid., pp. 549-553, "Japan and China: Treaty and additional agreement relating to Manchuria", or the Treaty of Peking, signed on December 12, 1905.

[2]Chang Pe, Jih-pen tui hua chih chiao-tung chin-lueh [The Japanese aggression to Chinese communication] (Shanghai: Commercial Press, 1931), pp. 55-57.

[3]For example, the joint construction of the Chilin-Ch'angch'un line with China from 1909. See Chang Pe, Jih-pen tui hua chih chiao-tung chin-lueh, pp. 57-61.

[4]This is well manifested in the widespread and popular demands for reform and the sporadic attacks upon foreigners in treaty ports throughout the country, the former culminating in the "One Hundred Day Reform" of 1898 led by K'ang Yu-wei and Liang Ch'i-ch'ao, and the latter in the Yi Ho Tuan Movement of 1900.

[5]Mongton Chih Hsu, "Railway Problems in China," pp. 89-102.

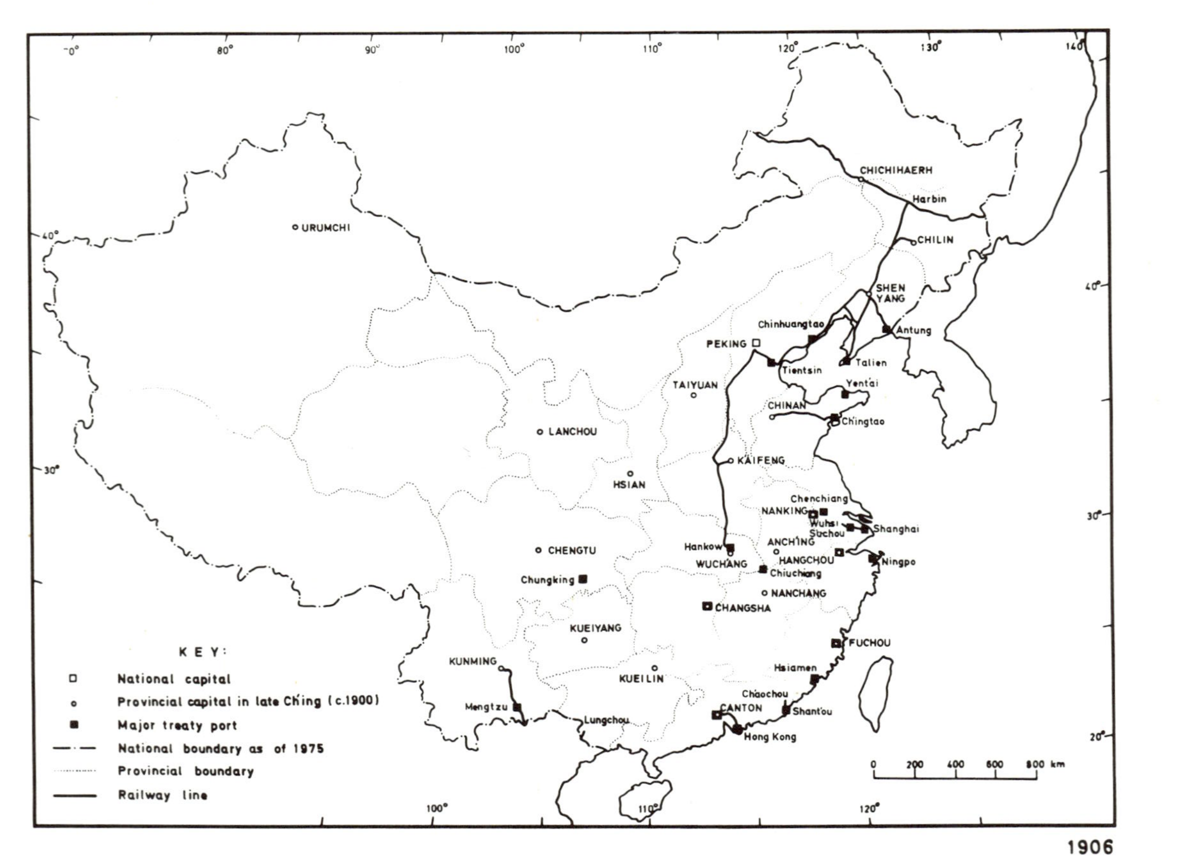

Fig. 3. Railways in China, 1906

For sources, see appendix D.

Sanshui line in 1902, construction of the Ch'aochou-Shant'ou railway was approved in 1903,[1] and many others were planned, including the Hsin-Ning line in Kuangtung and the inter-provincial Canton-Hsiamen line.[2] At the same time, almost every province, with the encouragement of provincial authorities and the enthusiastic support of local gentry and overseas Chinese, began to plan or build provincial and private railways. Between 1903 and 1907, no less than twenty-six provincial railway companies were formed[3] and, in spite of the prevailing economic conditions, were fairly successful in capitalization, especially in the coastal provinces, such as Chechiang, Kuangtung, Chiangsu, and Fuchien (table 1). In less than ten years time, by 1911, some 650 kilometers of railway lines, or roughly 7.2% of the national total, had been constructed (table 2) by private and provincial enterprises, and the annual average construction of some 70 kiolmeters compared favourably with the progress made by Chinese official efforts during the period before 1898 (table 3).

Nationalization and the Fall of the Ch'ing Dynasty

The decision to nationalize all trunk lines in 1911 was prompted by the desire to centralise planning and to speed up the development of railways in China. Shortly after the emergence of the private and provincial railway companies, the Bureau of Trade memorialised in 1906 that,

> . . . since this Bureau coordinates all road administration, it should develop a comprehensive plan . . . Thereafter applications from provincial authority and gentry for railway construction could be called for in accordance with priorities set and routing compiled by this bureau . . .[4]

Two years later, the Kuang Hsu Emperor complained that the slow progress of the private lines by provincial gentry and authorities not only incurred great losses but also interfered with the development of communications.[5] Serious financial problems and malpractice were found in some provincial railway companies,

---

[1]Fu Yu-ch'eng, Chung-kuo chin-t'ai tieh-lu shih tzu-liao, 1863-1911, pp. 929-931.

[2]Ibid., pp. 923-963.

[3]Ibid., pp. 1147-48.

[4]Chiao-tung shih lu-cheng pien, pp. 864-5.

[5]Fu Yu-ch'eng, Chung-kuo chin-t'ai tieh-lu shih tzu-liao, 1863-1911, p. 1157.

TABLE 1

CAPITALIZATION OF PROVINCIAL RAILWAY COMPANIES
(in '000 dollars up to 1911)

Province	(1) Nominal Capital	(2) Paid up Capital	(3) % Capitalised (2)/(1)
Chechiang	6,000	9,250	154.2
Ssuch'uan	20,990	16,450	78.4
Kuangtung	20,000	15,130	75.7
Chiangsu	10,000	4,100	41.0
Hunan	20,000	6,520	32.6
Heilungchiang	1,400	450	32.1
Chianghsi	6,990	2,190	31.3
Fuchien	6,000	1,700	28.3
Shanhsi	2,800	320	11.4
Hupei[a]	36,000	2,120	5.1
Anhui	–	1,240	?
Shenhsi			
Loyang-T'ungkuan line	15,000	300	2.0
Hsian-T'ungkuan line	5,590	–	0
Kuanghsi	30,000	–	0
Hunan (branch line)	11,890	–	0
Yunnan	–	–	–

SOURCE: Fu Yu-cheng, Chung-kuo chin-t'ai tieh-lu shih tzu-liao, 1863-1911, pp. 1149-50.

[a]Sections of the Kuangtung-Hupei-Ssuch'uan railway.

TABLE 2

RAILWAY CONSTRUCTION BY PRIVATE AND PROVINCIAL COMPANIES
(Up to 1911)

Line	length in km	Remark
Private[a]		
Canton-Sanshui	48.92	completed in 1902
Ch'aochou-Shant'ou	39.10	" " 1908
Chang-Hsia	28.00	" " 1910

TABLE 2-Continued

Line	length in km	Remark
Hsin-Ning	109.60	from Toushan to Peichieh, Kuangtung, completed in 1913
Total private	197.62	
Provincial[b]		
Chechiang	148.80	up to 1909 only
Fuchien	17.00	- do -
Chiangsu	55.60	- do -
Heilungchiang	25.30	- do -
Chianghsi	40.00	
Anhui	—	9 km road bed completed
Hunan	50.30	
Kuangtung	85.00	
Ssuch'uan	—	of 280 km from Ichang to Kueichou, 30 km completed for construction transportation.
Total provincial	422.00	
Grand total	647.62	

[a]Yen Chung-p'ing, et al., Chung-kuo chin-tai ching-chi-shih t'ung-chi tzu-liao hsuan-chi [Selection of statistical materials on Chinese modern economic history, 1760-1949] (Peking: Ko Hsueh chu-pan-she, 1955), pp. 172-177, table 1.

[b]Fu Yu-cheng, Chung-kuo chin-t'ai tieh-lu shih tzu-liao, 1863-1911, pp. 1150-1151.

resulting in their liquidation or nationalization as early as 1908.[1] Finally, on May 9, 1911, an Imperial Edict was issued proclaiming the nationalization of all trunk lines.[2]

Ironically, while private and provincial companies which largely emerged in defiance of foreign concessions and foreign loans were being nationalized, the government entered into more loan agreements with the powers thereby giving up more rights.[3] Railway

---

[1]Fu Yu-ch'eng, Chung-kuo chin-t'ai tieh-lu shih tzu-liao, 1863-1911, pp. 1161-62.

[2]Ibid., p. 1236, and Mongton Chih Hsu, "Railway Problems in China," pp. 116-7.

[3]Ch'en Hui, Chung-kuo tieh-lu wen-t'i pp. 65-70.

associations which had been formed with the explicit objective to retrieve railway and other concession rights therefore soon joined forces with the local gentry to protect local lines. Protests were particularly widespread in Hunan, Kuangtung, and Ssuch'uan,[1] and finally led to the success of the October 1911 Revolution and the downfall of the Ch'ing Dynasty.

Patterns and Functions of Railways in Imperial China

Network Patterns and Characteristics

By the end of the Ch'ing dynasty, some 9,000 kilometers of main line plus 560 kilometers of branch lines had been constructed in mainland China (table 3). The period of most rapid development, with an average annual construction of some 800 kilometers of main and branch lines, occurred between 1899 and 1905, at the height of the foreign scramble.

TABLE 3

RAILWAY CONSTRUCTION IN CHINA[a] 1876-1911
(in kilometers)

Period	Total at end of period	Construction during period	Average yearly construction
	Main Line		
1876	15.00[b]	—	—
1877-1894	287.27	287.27	15.96
1895-1898	507.27	220.00	55.00
1899-1905	5,775.64	5,268.37	752.63
1906-1911	9,016.10	3,240.37	540.06
	Branch Line		
1876	—	—	—
1877-1894	—	—	—
1895-1898	31.27	31.27	7.82
1899-1905	417.97	386.70	55.24
1905-1911	561.75	143.78	23.96

SOURCE: Calculated from Yen Chung-p'ing, Chung-kuo chin-tai ching-chi-shih t'ung-chi tzu-liao hsuan-chi, 1955, tables 1 and 2, pp. 172-179.

[a]Railway in Taiwan excluded.

[b]This line was removed in the following year (1877) and is not included in subsequent statistics.

---

[1]Tai Chih-li, ed., Ssuch'uan pao-lu yun-tung shih-liao [Historical materials on Ssuch'uan Railway Preservation Movement] (Peking: 1959).

It will be seen (figure 4) that railway development during this period can be characterized by several distinct patterns. First, there was the semi-colonial pattern with lines typically penetrating inland from Treaty ports or foreign territories, as has been best described by Fisher and later by Gould, and Taaffe et al. as the Phase II of a colonial network.[1] However, unlike true colonial systems in Africa, India, and elsewhere, the semi-colonial pattern in China was owned or controlled not by one single but by a multitude of colonial powers. These powers, through loans rather than military imposition, planned, constructed, and controlled a number of regional networks, designed primarily to strengthen and extend their respective "spheres of influence". Since most railway lines were constructed in different gauges, they were seldom connected with each other nor did they form an integrated national network. Typically, the system was import- rather than export-oriented.

Second, there was the nationalistic pattern with lines radiating from the capital. Essentially conceived as a means of national defence and territorial integration, these lines were aimed either at the vulnerable frontier or coastal areas, or at connecting important administrative centers in the core areas. Third, there was the shortlived and minor corporate pattern developed by free enterprises and local capitalization, particularly in the richer coastal provinces. Since these lines were built without coordination or overall planning, they were mostly short, scattered, and inefficient local lines.

On the whole, the development of modern transportation in China did not lead to the emergence of high-priority corridors, nor the increased primacy of major port cities. Indeed, as table 4 shows, the dominance of Shanghai and to some extent Canton, in handling import and export trade declined rather than improved with the introduction of railways since the 1880's. Ports on the Yangtzu River and other minor ports apparently unconnected by rails continued to retain their relative role, and after 1905 railway terminals distinctly representing interests of different colonial powers such as Ch'ingtao, Talien, Harbin, and Antung, emerged to compete for the same trade. This was in direct contrast to

---

[1] C. A. Fisher, "The Railway Geography of British Malaya,"; Peter R. Gould, The Development of the Transportation Pattern in Ghana; and Edward J. Taaffe, et al., "Transport Expansion in Underdeveloped Countries: A Comparative Analysis."

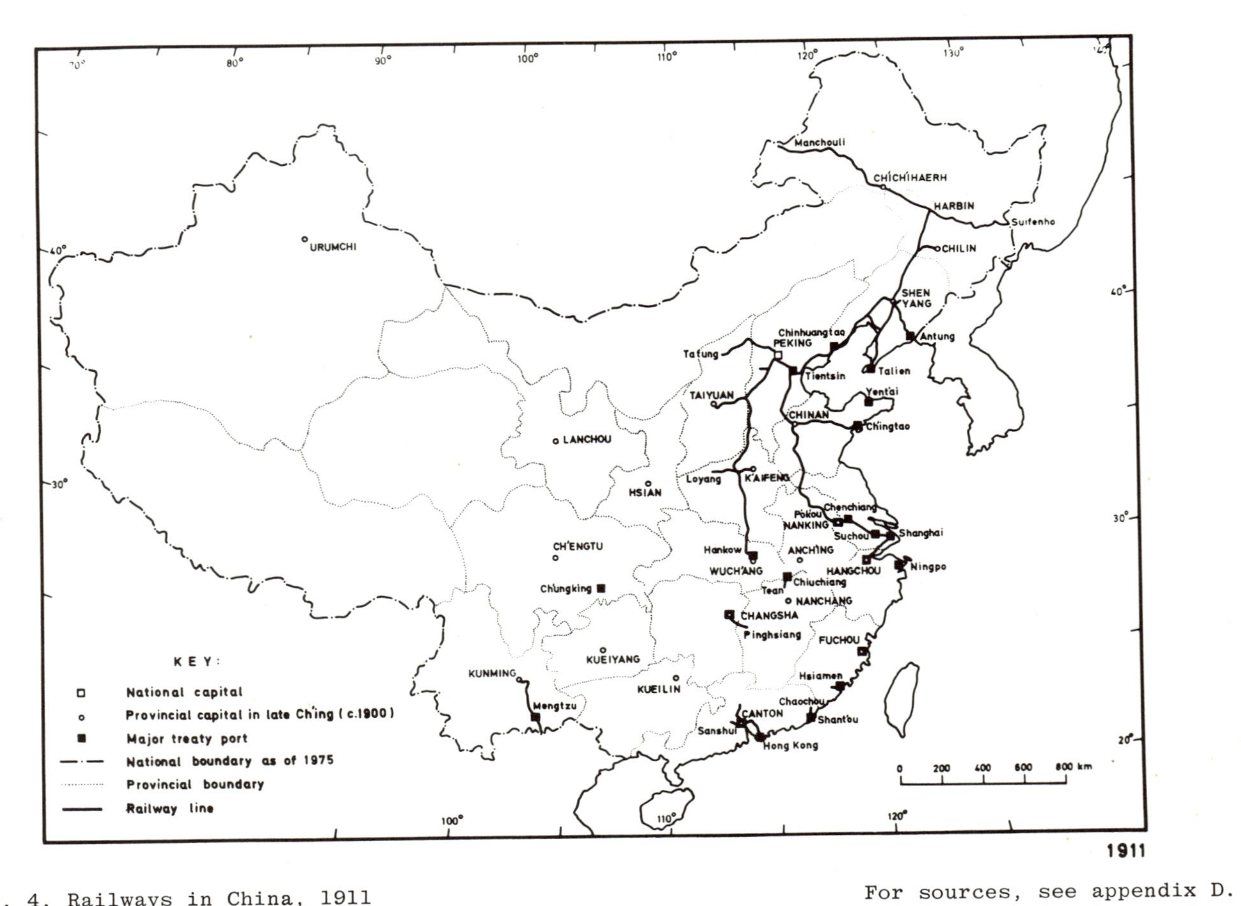

Fig. 4. Railways in China, 1911

For sources, see appendix D.

TABLE 4

CHINA'S FOREIGN TRADE BY PRINCIPAL PORTS, 1870-1948
(in percentage)

Year	Shanghai	Canton	Tientsin	Hankow	Ch'ingtao	Other Yangtzu Ports	Talien	Harbin	Total
1870	72.25	15.23	1.70	1.93	—	1.93	—	—	93.04
1875	56.08	11.81	2.92	4.21	—	4.43	—	—	79.45
1880	58.68	10.02	3.46	4.88	—	5.06	—	—	82.10
1885	56.83	10.58	3.33	4.92	—	4.93	—	—	80.59
1890	46.21	12.12	3.02	2.92	—	2.74	—	—	82.12
1895	53.60	10.84	4.54	1.71	—	2.17	—	—	86.91
1900	55.16	8.81	1.30	1.88	0.05	2.48	—	—	80.93
1905	54.27	9.42	6.21	5.33	1.01	6.34	—	—	88.17
1910	44.32	10.26	4.57	4.35	2.38	5.41	4.61	4.35	86.49
1915	45.76	7.81	6.59	5.06	0.86	6.20	6.93	3.82	89.78
1920	44.31	7.04	6.76	4.16	3.07	6.13	13.99	1.41	93.63
1925	42.81	5.24	8.54	4.87	3.93	6.27	12.02	3.21	93.82
1930	45.02	5.06	8.30	2.06	4.04	3.49	14.51	3.04	91.11
1935	53.27	4.93	11.79	3.06	6.67	5.47	—	—	90.27
1940	53.24	0.75	20.26	0.002	8.05	0.34	—	—	85.60
1946	80.52	4.36	4.80	0.07	1.75	0.16	—	—	94.07
1947	69.44	5.10	5.62	0.03	1.08	0.01	—	—	91.48
1948	74.58	7.55	3.57	3.74	0.98	0.03	—	—	95.39

SOURCE: Calculated from Hsiao Liang-lin, China's Foreign Trade Statistics, 1864-1949 (Cambridge: Harvard University Press, 1974), tables 1 and 7a, pp. 22-25, 168-178.

experiences in Africa and other colonial territories.[1]

The Economics and Functions of Chinese Railways

China's investment in her railway program may be gauged by the amount of foreign loans for railway construction. Table 5, which summarises foreign railway loans to China from 1887 to 1911, shows that up to 1911, China accumulated actual debts totalling more than 510 million yuan out of a total contracted amount of 558 million for railway construction, whereas her average annual indebtedness jumped from less than a quarter of a million for the period 1887-1894 to more than 47 million for the period 1906-1911. In terms of China's total annual export capacity of between 167 million and 570 million during the period 1891-1911,[2] her annual railway debts alone were worth between 7 and 11 per cent roughly, while the entire railway program including government and private capitalization, expenditures on operation, maintenance, etc. at least double this value.[3] Since imported railway materials absorbed 50 to 90 per cent of capital

TABLE 5

FOREIGN RAILWAY LOANS TO CHINA, 1887-1911

	Contract Loan Amount (Yuan)	Actual Debt (Yuan)	Average Actual Debt p.a. (Yuan)
before 1887	0	0	0
1887-1894	4,302,098	1,693,007	211,626
1895-1898	71,535,698	75,383,768	18,845,942
1899-1905	178,521,134	153,615,000	21,945,500
1906-1911	304,009,635	283,180,047	47,196,675
Total	558,368,565	510,691,775	20,427,671

SOURCE: Calculated from Yen Chung-p'ing, et al., Chung-kuo chin-tai ching-chi-shih t'ung-chi tzu-liao hsuan-chi, table 7, pp. 190-193.

[1]Peter R. Gould, The Development of the Transportation Pattern in Ghana; and Edward J. Taaffe, et al., "Transport Expansion in Underdeveloped Countries: A Comparative Analysis."

[2]Yen Chung-ping, Chung-kuo chin-tai ching-chi-shih t'ung-chi tzu-liao hsuan-chi, p. 64, table 6.

[3]Arthur Rosenbaum, "Railway Enterprise and Economic Development," pp. 227-273.

expenditure on industrial goods for railway construction,[1] the Chinese railway program, instead of contributing to economic growth in general and industrialization in particular, was in fact a considerable drain on limited resources.

Moreover, railways built with foreign loans or control were frequently more costly, financially unsound, and commercially not necessarily the most desirable. According to Rosenbaum, Chinese-built railways without much loan burden were invariably less costly[2] (table 6). The Shanghai-Nanking and Tientsin-Puk'ou lines, for example, even after deducting for financial accounts, rolling stock and mechanical plant, has been found 50% more expensive than either the Imperial Railway of North China or the Peking-Suiyuan line, both constructed fairly independently by China. As a result of high brokerage and interest rates, excessive over-capitalization from the very beginning left many railway lines in serious financial difficulty for a long time. For instance, the lavishly equipped Shanghai-Nanking Railway had interest charges equivalent to nearly three times its operating profits in 1909, and the Tientsin-Puk'ou Railway's profit could hardly cover charges for interest and amortization.[3] Similarly, the Chinese Eastern Railway, due to over-capitalization and the fact that it was built long before demand, suffered annual losses from its opening in 1903 for 11 years totalling more than 174 million roubles.[4] Furthermore, foreign-financed railways incurred large expenses on security, legal, political, and special and secret services, which the lines could hardly afford.[5]

---

[1]Ibid.

[2]Similar conclusion had been arrived at by G. B. Rea, who found in 1909 that "the average cost of railways built under foreign loan agreements is $45,000 gold per mile, while serviceable railways can be constructed and equipped for $20,000 to $30,000 gold per mile when built without foreign interference." Rea therefore concluded that China could more than double her railway building if unmolested in the administration of her affairs. See Mongton Chih Hsu, "Railway Problems in China," p. 179.

[3]Arthur Rosenbaum, "Railway Enterprise and Economic Development."

[4]Ch'en Hui, Ching-kuo t'ieh-lu wen-t'i, p. 59. It should be noted that the line was built for military purpose and was cut off from the Northeast Plain since 1905 when the South Manchurian Railway was transferred to Japan.

[5]Ibid.

TABLE 6

AVERAGE CONSTRUCTION COSTS OF RAILWAYS IN CHINA, (dollars per km, 1916[a])

Railway	total capital costs	costs (excluding financial accounts)	Costs (excluding financial accounts, rolling stock, and mechanical works)
Imperial Railway of North China	61,896	61,416	42,355
Peking-Hankow	76,817	62,414	48,993
Peking-Suiyuan	56,460	53,926	40,720
Shanghai-Nanking	93,196	83,021	66,223
Tientsin-Puk'ou	90,169	78,552	67,355

SOURCE: Arthur Rosenbaum, "Railway Enterprise and Economic Development," pp. 227-273; and Ministry of Communications, Statistics of Government Railways (1917), table VII.

[a]With the exception of the segment from Changchiak'ou to Kueisui of the Peking-Suiyuan line, all other lines were completed by or before 1911.

In sharp contrast were the Peking-Shanhaikuan line and the South Manchurian Railway, both of which were highly profitable, the latter particularly after its transfer to Japan in 1905. The former was due largely to its freedom from loan interests and efficient and inexpensive management,[1] the latter to the availability of feeder lines[2] and the establishment of a large number of related enterprises by the Japanese South Manchuria Railway Company that helped take full advantage of the forward linkage effects of modern transportation.

On the other hand, railway loans to China were clearly most profitable to the powers, as they were credited at only 90 per cent or less and carried an interest rate of 8 per cent or more. Moreover, through these loans, the powers gained not only control[3] of

[1]Arthur Rosenbaum, "Railway Enterprise and Economic Development."

[2]Ibid.; and Ch'en Hui, Chung-kuo t'ieh-lu wen-t'i.

[3]Foreign control included rights of construction, management, recruitment of staff, operation, accounting, policing, and even of contracting additional loans. See Yen Chung-ping, et al., Chung-kuo chin-tai ching-chi-shih t'ung-chi tzu-liao hsuan-chi, tables 4 and 5, pp. 184-189; and appendix A.

the lines concerned, but numerous privileges such as mining and property rights along the lines, preferential traffic concessions, etc. Thus, through direct investment and loan, the powers' control of Chinese railways rapidly increased from about 79 per cent of the total railway network in 1893 to 93 per cent by 1911 (table 7).

TABLE 7

FOREIGN CONTROL OF CHINESE RAILWAYS
1894 and 1911

	1894		1911[a]	
	Km.	%	Km.	%
Independent	77.0	21.1	665.62	6.9
Foreign Controlled				
Direct investment[b]	–	–	3,759.70	39.1
Controlled investment[c]	287.27	78.9	5,192.78	54.0
Total	364.27	100.0	9,618.10	100.0

SOURCE: Yen Chung-p'ing, et. al., Chung-kuo chin-tai ching-chi-shih t'ung-chi tzu-liao hsuan-chi, table 6, p. 190.

[a]Including railways in Taiwan.

[b]For example, Russia in the Chinese Eastern Railway, Japan in the South Manchurian Railways, etc.

[c]That is, through loans concessions.

Summary of Findings

Throughout the last sixty years or so of the Ch'ing dynasty the obsessive interest of the leading mandarins and intellectuals in finding a means for modernization in general and for national defence in particular was hardly a subjective matter. Ever since the Eighteenth Century, the Chinese national territory was shrinking rapidly,[1] and many defeats in the early Nineteenth Century convinced

---

[1]From the 1690's onwards, China's national space, excluding areas of all dominions and protectorates, contracted from at least 13 million square kilometers to roughly 11.165 million square kilometers by 1911 under the Republican Government, and to only 9.6 million square kilometers by 1949 on the birth of the People's Republic of China. See Fang Ch'iu-wei, Chung-kuo pien-chiang wen-ti shih-chiang [Ten lectures on China's frontier problems] (Shanghai: Motor Press, 1937), table 2, between pp. 4 and 5; and Chung-hua jen-min kung-ho-kuo fen-sheng ti-t'u chi [Atlas of provinces of the People's Republic of China] (Peking: Ti-t'u ch'u-pan-she, 1974), p.7.

her that she must strengthen herself along Western lines, specifically through the adoption of Western technology and industry.[1] The railway was only one of such means and was aspired to provide the best means for national defence and territorial integration.[2] That these aspirations failed of realisation due to the absence of a body of universally accepted ideology in Imperial China was obvious, though the potentials of the railway were finally fully accepted.

First, the ideological split on the role of the railway seriously delayed its introduction. It follows that the choice of location for the first railways too was seriously restricted. The advocates of those lines not only had to avoid building a line that would compete with the livelihood of a large traditional sector, but also had to demonstrate its need. The first consideration practically meant a line of least resistance away from the most densely populated areas, or the Chinese ecumene, where its successful introduction would have been better guaranteed. The second consideration pointed to strategic lines the need for which was readily demonstrable.

Second, when the strategic role of the railway was more acceptable, there was a lack of agreement as to which constituted the most urgently needed and most effective lines. The concept of territorial integration was violated when foreign powers were allowed to share the construction and the control of the defensive system, which finally led to conflicts in an area which the Imperial government aspired most to safeguard.

Third, the obsession with the strategic role of the railway rendered it from the beginning a central government concern. Li Hung-chang never wanted to entrust the Chinese merchant class with the entire enterprise, and therefore devised the system called kuan-tu shang-pan (government supervision, merchant management). Even when Chang Chih-tung took over, he conceded only kuan-shang ho pan (government-merchant co-operation). This begot the constant conflicts between political and commercial interests and resulted in a dearth of local capital.[3] By the time that the provincial gentries

---

[1]Chester Tan, Chinese Political Thought in the Twentieth Century, (Garden City: Doubleday, 1971), pp. 7-8.

[2]Other means China had already adopted included steel works, modern arsenals, a navy, the telegraph, and ocean navigation.

[3]Indeed, there was no lack of Chinese capital. Since the merchant class was suspicious of the bureaucrats, they either withheld their capital or contributed it to foreign loans to China for interests and security purpose. See Fu Yu-ch'eng, Chung-kuo chin-t'ai tieh-lu shih tzu-liao, 1863-1911, p. 163.

were roused to build private (min ying) railways, most of the important trunk lines had already been concessioned, and the revolutionary tide was rising with nationalism.

However, the most dominant factor in the process was no doubt the interference of the foreign powers whose objectives in railway building in China were hardly entirely commercial, for capitalist expansion required not only market but also territory. Thus China's cautious attitude toward the functions of the railway and her resistance to foreign penetration were seen as against the "spirit of the age". Moreover,

> It has been evident . . . that . . . foreigners would and must have extended freedom of intercourse with the interior of China; . . . if the country were not opened up from within by the Chinese themselves, the spirit of the age, . . . would be too much for her; . . . what was not conceded . . . would be wrung from her by the force of circumstance, culminating in the dismemberment of the Empire . . .[1]

Eventually, an extremely fragmented railway system was constructed, since the gauge as well as the location of each line were not infrequently determined by the foreign powers. These lines not only penetrated into the interior from the treaty ports extending the markets of foreign goods and generating profits for the foreign powers, but also secured and expanded their spheres of influence. The establishment of spheres of influence, thereby rendering China a semi-colonial country, proved the bankruptcy of the railway policies of China in the face of capitalist expansion.

---

[1]Archibald R. Colquhoun, China in Transformation (London: Harper, 1898), p. 94.

CHAPTER III

RAILWAY PATTERNS, NATIONALISM, AND THE DEFEAT OF THE REPUBLIC OF CHINA

Railway Development and National Aspirations

The Sun Yat-sen Doctrine and His Development Plans

The long period of foreign encroachment upon the Chinese national space saw not only the evolution of responsive ideologies and aspirations, but also the experimentation of practical policies for their attainment. The former is evidenced by the long debate over the question of "Westernization" between the reformists and the conservatives from the last half of the Nineteenth Century onward, culminating in the prevalent formula of "Chinese learning for substance, Western learning for function" by the turn of the century.[1] The latter is exemplified by the adoption of Western science and technology, such as the establishment of modern navy, arsenals, merchant shipping, and the railway system. The attitude toward the political and military role of the railway underlined the evolution of railway policies from kuan-tu shang-pan (government supervision, merchant management), to kuan-shang ho-pan (government-merchant co-operation), to limited min-ying (private enterprise), and finally to nationalization which contributed to the down fall of the Imperial Government.

Sun Yat-sen, leading the nationalist revolution and seeing that national pride was facing bankruptcy, emphasized that "revolution must begin with the revolution of mind,"[2] and propounded as the

---

[1]According to Chester Tan, this was "A syncretic view intended to strike a balance between reformism and conservatism, was offered by Chang Chih-tung (1837-1909)," who defended traditional institution and was loyal to the Manchu Monarchy. See Chester Tan, Chinese Political Thought in the Twentieth Century, pp. 9-10.

[2]Hsiung Heng-ling and Ho Te-chuan, Kuo-fu hsueh-shuo yu chung-kuo t'ieh-lu [Dr. Sun Yat-sen's doctrine and Chinese railways] (Taipei: Railway Nationalist Party Committee, 1966), vol. 1, p. 5.

first principle the principle of nationalism in his san-min chu-i (the three principles of the people) to arouse nationalism. His popularly known democratic ideology was based neither on the theory of natural right nor on the contract theory of the state, for he not only was opposed to individualism, but also regarded the state as an organized body for mutual assistance rather than a conglomeration of individuals for the protection of each one's rights.[1] In the same vein, the objective of the principle of people's livelihood was to "restrict capitalism and to even distribution."[2] Like other national leaders, he aspired to treat railways as an important means of national reconstruction. As early as 1894, he petitioned Li Hung-chang on national salvation programs and, in terms of railway development, criticised the policy of giving priority to the extra-mural line[3] rather than lines in the most prosperous areas.[4] He too held that railways should be nationalised,[5] that foreign loans should be employed for rapid development and long-term benefit and, after the fall of the Ch'ing government, that soldiers should be demobilised and transformed into railroad builders for the peaceful unification of the country.[6] After the establishment of the Republic of China, Sun, at a press conference in Peking in 1912, pointed out that the presence of Russia in northern Manchuria and Mongolia, of

---

[1]Chester Tan, Chinese Political Thought in the Twentieth Century, pp. 120-126.

[2]Furthermore, in his conception of san-min chu-i or three principles of the people, he made anti-imperialism a very important part of the principle of nationalism, especially Chinese nationalism; the idea of one-party rule in a guided democracy, with the Kuomintang or Nationalist Party above the state, the backbone of his principle of democracy; and popular rather than monopolistic ownership the essence of the principle of people's livelihood. See Shan Chuan Leng and Norman D. Palmer, Sun Yat-sen and Communism (London: Thames and Hudson, 1961), pp. 93-94.

[3]That is, essentially, the Peking-Shenyang line consisting of the original segments developed by Li Hung-chang and discussed in chapter 2, section 1 above.

[4]Hsiung Heng-ling, et al., Kuo-fu hsueh-shuo yu chung-kuo t'ieh-lu, pp. 6-7.

[5]Ibid., pp. 9-10. However, Sun was opposed to nationalization under the Imperial Government on the grounds that the monarchical government, not being one popularly elected, was even more monopolistic in nature than a few capitalists.

[6]Ibid., pp. 59-63.

Japan in southern Manchuria, and of Britain in Tibet, was due to weakness in defence and difficulty in communication and transportation. He therefore emphasized that the most urgent and the only task then was railway-building, for on it was hinged the existence of the Republic.[1]

That railway development was crucial to his plans for national reconstruction can be seen from his devotion to railway problems after resigning from the Presidency in 1912.[2] After the First World War, his long-advocated railway plan was incorporated into his program, The International Development of China, presented to international consortia and foreign financiers in the hope of capturing the surplus materials and production capacity after the war. Despite the four principles he explicitly adopted for the routing, for example, that of choosing the most remunerative route to attract foreign capital, etc., the plan was seen by his successor, Chiang Kai-shek, as one for national defence, for

> . . . in terms of railway construction, there must exist port termini. But all the best ports . . . along the coast . . . are constrained by unequal treaties . . . (thus) the location of the three ports . . . in the plan is not absolutely ideal, but is constrained by circumstance . . . they may be called . . . main ports . . . or fishing ports, but are in fact all naval bases; whereas all railway centers and termini are in fact strategic . . . points in national defence . . .[3]

No doubt Sun's blueprint for national reconstruction through railway development, in which he saw the means for frontier defence, emigration of population, national investment and the opening up of the interior, continued to be one of the chief aspirations of the Nationalist government in the next several decades.[4]

---

[1]Wang Kuang, Kai kuo liu-shih nien chiao-tung shih lun [Sixty years of communications history of the Republic] (Taipei: By the Author, 1971), pp. 1-11.

[2]China Hand Book (Shanghai: North China Daily News and Herald, 1913), pp. 186-187.

[3]Hsiung Heng-ling, et al., Kuo-fu hsueh-shuo yu chung-kuo t'ieh-lu, pp. 122-3 and 128-9.

[4]Ibid. See also Chiang Kai-shek, China's Destiny (New York: Macmillan, 1947); Chang Kia-ngau, Chung-kuo t'ieh-tao chien-she; Ling Hung-hsun, Chung-kuo t'ieh-lu chih [A comprehensive survey of railway development in China] (Taipei: Hsieh Lung Printing, 1954), and Chung-kuo t'ieh-lu kai lun [On Chinese railways] (Taipei: National Publishing Bureau, 1950); and Wang Kuang, Kai kuo liu-shih nien chiao-tung shih lun, pp. 1-32.

Railway Strategies and Japanese Penetration

True to its aspirations, the Republic of China chose to continue the policies of nationalization of railways and of using foreign loans, both of which contributed to the downfall of the Ch'ing dynasty. That there was no resistance whatsoever to the nationalization of provincial railways in 1912 proved that the policy was widely accepted.[1] On the other hand, loan negotiations were immediately resumed for the construction of sections of the Hupei-Kuangtung-Ssuch'uan line, and the Lung-Hai railway.[2] In addition, new loans and contracts were vigorously pursued for the construction of railways in Shanshi, Anhui, Honan, Hupei, Hunan, Ssuch'uan, Kueichou and Kuangtung provinces (appendix A).[3] Though many of them never materialised due to the outbreak of the First World War, the location of these lines clearly showed that the new government was determined to strengthen the hold of the national capital at Nanking by focusing on lines south of the Yangtzu and to create a viable program that capitalized on the most developed areas.[4]

However, the establishment of the Republic coincided with Japan's rising ambition and influence in China, particularly in the Northeast.[5] The conversion to standard gauge of the South Manchurian and the Antung-Shenyang railways after 1905 meant that the lines and to some extent southern Manchuria were well integrated with Korea, then a Japanese colony.[6] By 1913, Japan agitated for the right to

---

[1]Ling Hung-hsun, Chung-kuo t'ieh-lu kai lun, p. 7.

[2]Ibid., p. 8.

[3]Ibid., specifically, they included the Frenco-Belgian loan for the T'ung-Ch'eng (Tait'ung-Fenglingtu) line; a Franco-Chinese Industrial Bank loan for the Chin-Yu (Chinchou-Chungking) line; the Chinese Central Railways Company loan for the Pu-Hsin (Puk'ou-Hsinyang) line, etc.

[4]Ibid., p. 11.

[5]It should also be noted that the European powers were then drawn away from Asia by the First World War (1914-19) and the Russian Revolution (1917).

[6]The South Manchurian Railway originally had a gauge of 3'6", and the Antung-Shenyang line one of 2'6". According to Chang K'oh-wei, Tung-pei kang jih ti tieh-lu cheng-tse [Railway policy for the resistance against Japanese in the Northeast] (n.p.: Liang-yu, 1913), p. 19, this action aroused the suspicion of Britain and the United States, to the extent that the Japanese henceafter had to employ other tactics, such as forcing loans on China, to extend her control of the Northeast through railway construction, etc.

provide loans to China for railway development in Manchuria.[1] Immediately after the outbreak of the First World War and taking advantage of the unstable position of Yuan Shih-k'ai, Japan forced on China the notorious Twenty-one Demands.[2] Consequently, several lines were completed with Japanese money, namely, the Ssuping-T'aonan line in 1923, the T'aonan-Angangch'i line in 1926, and the Chilin-Tunhua line in 1929.[3] Long before that, the military attache to the British Legation in Peking pointed out that the completion of these lines would make Japan's military position in Manchuria considerably stronger, and that,

> What Japan is aiming at is firstly to have through railway lines to each of the three places - Chilin, Ch'angch'un, and T'aonan fu, so as to be able to concentrate, if necessary, equal forces at each place, and secondly to have a line joining them so as to be able to transfer from one flank of their position to another.[4]

These lines, branching out from the South Manchurian Railway, not only soon became important feeders, but also formed integral parts of a grand program aimed at capturing traffic of the Chinese Eastern Railway and ultimately terminating at a port in northeastern Korea on the Japan Sea. In his memorial to the Japanese Emperor, Premier Tanaka stated the purpose of his railway program plainly,

> . . . the existing lines of the South Manchuria railways, being mostly economic-oriented, lacked circuits and were disadvantageous to mobilization and military transportation in war time. Henceforth the construction of the Manchurian-Mongolian great circuit lines must be military-oriented, so as to enclose the core area of Manchuria and Mongolia, to stop the growth of Chinese military, political, and economic strength . . .[5]

---

[1]Ling Hung-hsun, Chung-kuo t'ieh-lu kai lun, p. 8; and Chang K'oh-wei Tung-pei kang jih ti t'ieh-lu cheng-tse, p. 19. While the Japanese were pressing the Chinese Government for the construction of the following lines: (1) T'aonan-Ssupingchieh; (2) T'aonan-Jeho; (3) Hailung-Kaiyuan; and (4) Chilin-Huining, "the Russian authorities are considering what additional lines of railway in Northern Manchuria would provide adequate compensation." (F.O. 228/2494 Consul Sly to Sir J. Jordan, December 15, 1913).

[2]For full contents of the treaty, see Sino-Japanese Negotiations of 1915, pp. 2-8.

[3]Ling Hung-hsun, Chung-kuo t'ieh-lu kai lun, pp. 8-9.

[4]D. S. Robertson, "Notes on the strategic effect of projected railways in Manchuria," enclosure 3 in J. Jordan's despatch to Edward Grey. January 12, 1914. (F.O. 228/2494).

[5]Ch'en Hui, Chung-kuo t'ieh-lu wen-t'i, p. 49.

> the completion of the Chilin-Huining line means the success of our new continental policy . . . for then Siberia can be reached from Chongjin through Huining . . . By that time . . . (we can) realise the final stage of Emperor Meiji's plan to annex Manchuria and Mongolia . . . in order to conquer the Chinese mainland . . .[1]

Confronted with increasing Japanese pressure in the Northeast, China once again had to abandon an economic-oriented railway program and to follow the long tradition of defensive railway planning in the Northeast dating back to Li Hung-chang and Cheng Teh-chuan.[2] The most ambitious plan was proposed by Hsu Shih-ch'ang, Viceroy of the Northeast, in the early 1900's, to criss-cross the Chinese Eastern and South Manchurian railways, and to terminate in a completely Chinese controlled port to be built at Hulutao.[3] It was not until 1924, however, with the establishment of the Northeast Communication Council, that the policies of linking up provincial capitals in the region by Chinese-owned railways and the building of a new port at Hulutao were vigorously pursued. Three trunk lines were then planned: the eastern one from Shenyang via Chilin to Fuyuan, a frontier town at the confluence of the Ussuri and Heilungchiang; the western line from Tahushan via T'ungliao and Ch'ich'ihaerh to Aihui, on the Sino-Russian border; and the southern line from Hulutao via Chihfeng to Tolun.[4] All terminating at Hulutao by-passing the Japanese at Talien, and all to be constructed with Chinese capital, these lines were undoubtedly designed to cut off the lines of the South Manchurian Railway from their hinterlands,

---

[1]Chang K'oh-wei, Tung-pei kang jih ti t'ieh-lu cheng-tse, pp. 28-29. The authenticity of the Tanaka memorial, "Japan's Positive Policy in Manchuria," has long been debated. A secret note by the British legation accompanying an English version of the memorial to the British Foreign Office on July 27, 1931, stated that it was, ". . . a fair guide to what Japanese policy in Manchuria and Inner Mongolia." See F.O. 371/15446. In a recent work Shinobu Seizaburo, Nihon Gaiko-shi, 1853-1972 (Tokyo: Mainich Shinbun-sha, 1974) Vol. II, pp. 349-350, feels that the contents of the Memorial not only agreed with Japan's continental policy, but were testified later by the aggressive path of Japanese imperialism.

[2]That is, the Hsin-Ai railway which Cheng proposed in 1906. See Yuan Wen-chang, Tung-pei tieh-lu wen-ti [Railway problems in the Northeast] (Shanghai: Chunghua, 1932), p. 4

[3]They are, the Chin-Ai, Hsin-Ch'i, and Shen-Yen lines, first proposed by Hsu in 1908. See Yuan Wen-chang, Tung-pei tieh-lu wen-ti, pp. 4 and 92-96; and Harry L. Kingman, Effects of Chinese Nationalism upon Manchurian Railway Developments, 1925-1931 (Berkeley: University of California Press, 1952).

[4]Ch'en Hui, Chung-kuo tieh-lu wen-t'i, pp. 50-51; and Yuan Wen-chang, Tung-pei tieh-lu wen-ti, pp. 7-8.

and to emasculate the Chinese Eastern Railway.[1]

Japan had mixed reaction to the eastern trunk line, since the first sections of which from Shenyang via Hailung to Chilin formed, with the Chilin-Ch'angch'un line, a small circuit with the South Manchurian Railway and was part of Tanaka's plan. Having failed to gain control of its construction, Japan forced China to allow the operation of the section between Shenyang and Hailung since its completion in 1928 to be linked to the South Manchurian Railway.[2] The construction of the western trunk line, however, met with strong Japanese opposition, on the ground that the route constituted a parallel line to the South Manchurian Railway.[3]

With the establishment of the Nanking government in 1928, railway construction in the Northeast was more rapid. Thus, before the Japanese annexation of the Northeast, the eastern trunk line reached Chilin, the provincial capital, in 1929 linking up with the Chilin-Tunhua line which had been completed one year earlier; the western trunk line was completed well beyond Ch'ich'ihaerh reaching Keshan in Heilungchiang (fig. 5). The completion of these lines enabled the Chinese to implement a unified management of traffic from Tientsin through Tunhua on the east, and to Keshan on the west, without using the South Manchurian Railway if necessary.[4] In addition, a line north of the Ch'angch'un-Harbin segment of the Chinese Eastern Railways was built from Harbin northward to Hailun. This was to ultimately link up the western trunk line at Keshan, and obviously to divert freight from the South Manchurian Railway.

In short, almost half of China's railway construction during this period took place in the Northeast,[5] and more lines were constructed there with Chinese capital than elsewhere in China. It was the Chinese determination to re-integrate this area through a vigorous railway program that prompted Japanese occupation of the entire Northeast in 1931, though there were clearly other considerations

---

[1]Norton Ginsburg, "Manchurian Railway Development," Far Eastern Quarterly 8 (August 1949): 398-411.

[2]Yuan Wen-chang, Tung-pei tieh-lu wen-ti, pp. 8 and 18.

[3]Ibid., p. 9.

[4]Chin Shih-hsuan, ed., Chung-kuo tung-pei t'ieh-lu wen t'i hui-lun [China's northeastern railway problems] (Tientsin: Ta Kung Pao, 1932), pp. 70-76; and Tieh-lu yun-shu ching-yen tan [Experiences on railway transportation] (Chungking: Cheng Chung, 1943), p. 31.

[5]Between 1912 and 1931, 2184 km or 47.3% of a national total of 4618 km of railway construction were built in the Northeast. See Yen Chung-p'ing, et al., Chung-kuo chin-tai ching-chi-shih t'ung-chi tzu-liao hsuan-chi pp. 172-177, table 1.

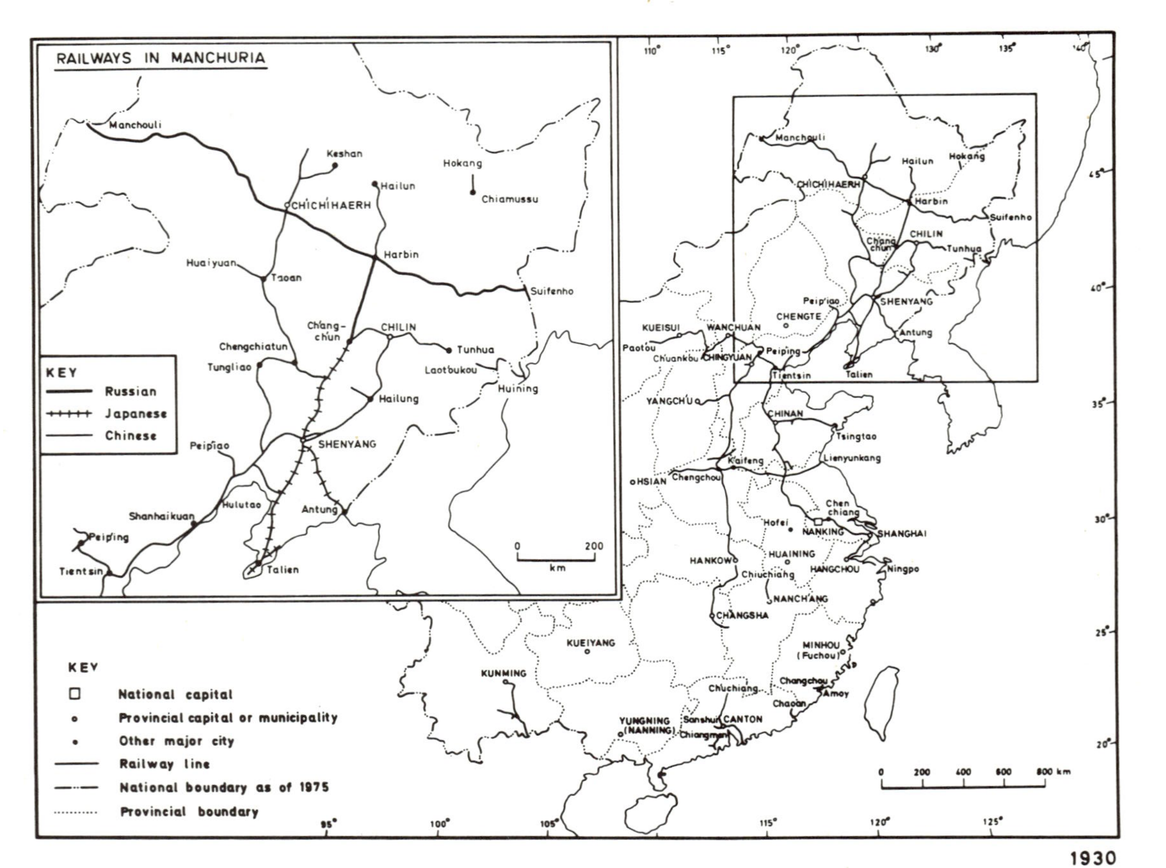

Fig. 5. Railways in China, 1930

For sources, see appendix D.

including the competition of Chinese railways to the South Manchurian Railway Company, the Japanese agent of control in the area, and the large-scale Chinese immigration into the area.[1]

Railway Development and a Divided Space-Polity

The period between the formal establishment of the Manchurian government Manchoukuo in 1932 and the outbreak of the Sino-Japanese war in 1937 saw the most intensive railway program in the history of Chinese railroading.[2] In both parts of Chinese territories, railway construction was vigorously pushed ahead, but was clearly differently aspired.

Railway Strategies and Shrunken National Space

To China, the loss of Manchuria, only eight years after the unilateral declaration of independence of Outer Mongolia in 1924,[3] meant not only a much shrunken national space, but also the complete bankruptcy of a railway strategy, developed over several decades, to directly counteract foreign encroachment in the contested areas of the Chinese national space.

Indeed, by 1929, the imminent threat was only too obvious, and the Central Government, largely following Sun's railway program, formally adopted the policies of (1) giving priority to railway construction in the Southwest over the Southeast; (2) developing a Northwest network to connect with that of the Southwest and the last

---

[1]Norton Ginsburg, "Manchurian Railway Development." See also Michael H. Hunt, Frontier Defense and the Open Door (New Haven: Yale University Press, 1973), p. 246.

[2]See table 9 below.

[3]Indeed the process of dismemberment of China had a long history. Outer Mongolia's agitation for independence, supported by Russia, dates back at least to 1911, and Tibet's, supported by Britain, to 1913. In 1924, Outer Mongolia passed a new constitution and unilaterally reiterated its independence. However, even Russia recognised in the same year that Outer Mongolia was part of China, whereas China in the 1931 Congress only agreed to the principle of self-government for Outer Mongolia. See Ssu Mu, Chung-Kuo pian chiang wen-ti chiang-hua [Talks on China's frontier problems] (Shanghai: Sheng-ho shu-tien, 1937), pp. 78-81; Theodore Herman, "Group Values Toward the National Space," pp. 299-329; and J. B. R. Whitney, China: Area, Administration, and Nation Building, pp.38-41.

line of defence; and (3) developing inter-provincial trunk lines.[1] By 1936, the Government's five-year railway plan included the construction of the following lines: Hsiangt'an-Kueiyang line, Paochi-Ch'engtu line, Ch'engtu-Chungking line, Hsiang-Kuei (Hunan-Kuanghsi) line, and many others.[2] It was quite clear that the policies shifted from a frontal approach to one that emphasised the strengthening of China Proper, or the Chinese ecumene, in anticipation of large-scale foreign invasion. Thus, after the completion of the Hankow-Canton and the Lung-Hai lines, the north-south and east-west trunk lines respectively, both having been started a long time before, the construction of important lines reaching into Central China such as the Tat'ung-Fenglingtu line, the Hangchou-Nanch'ang line, etc. was undertaken during the period. Other lines, such as the Nanking-Kueichi line forming a small circuit with the Hangchou-Nanch'ang line and strengthening the position of the capital, and the Ch'angsha-Kueiyang line had been started. The construction of the former stopped in November 1937, while the completed 175 kilometers of the latter were removed in 1939 for the construction of the Liuchou-Kueiyang line, after the outbreak of the Sino-Japanese war (fig. 6).[3] The construction of railways during this period not only had to take the possible Japanese invasion into consideration, according to Chang Kia-ngau, Minister of Railways, but also,

> Certain preparatory measures were taken by the railroads for the impending crisis, such as the extension of station facilities, precautions against air raids, storage of indispensable materials, and military training of the railroad staff.[4]

The Suchou-Chiahsing railway provided the best illustration of such railway policies. Soon after the so-called first Shanghai-Wusung battle of January 28, 1932 between China and Japan the Chinese government realised the importance of goods traffic between Nanking, the capital, and Hangchou, particularly after the expected completion

---

[1]Shih-nien lai chih chung-kuo ching-chi chien-she [Ten years of economic reconstruction in China] (China: Chung-kuo kuo-min tang, 1937), pp. 107-8. See also Hsiung Heng-ling, Kuo-fu hsueh-shuo yu chung-kuo t'ieh-lu pp. 156-57 and 158.

[2]Ling Hung-hsun, Chung kuo t'ieh-lu kai lun, pp. 11-12.

[3]Ibid., pp. 13-15.

[4]Chang Kia-ngau, China's Struggle for Railroad Development, pp. 182-83.

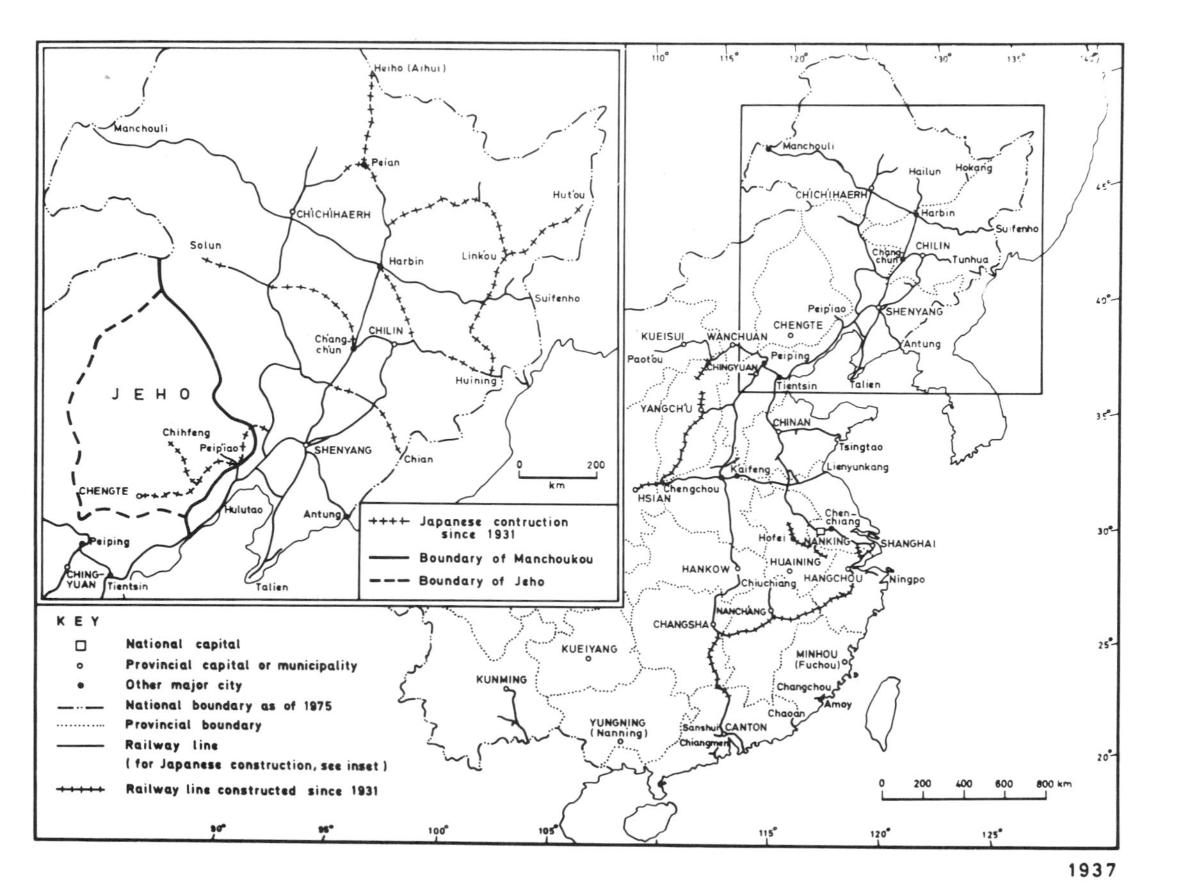

Fig. 6. Railways in China, 1937

For sources, see appendix D.

of the Hangchou-Nanch'ang line. Thus the Suchou-Chiahsing line bypassing Shanghai, the stronghold of foreign powers, was hurriedly completed in July 1936, shortening the distance between Nanking and Hangchou by about three hours. Moreover, railway staff were immediately given training in all aspects of war-time transportation. When the Sino-Japanese War spread to Shanghai on August 13, 1937, this line enabled China to mobilise its armies between Nanking and Hangchou freely, though it was soon completely destroyed by heavy bombing.[1]

Japanese Railway Strategies in Manchuria

Soon after the September 18, 1931 incident, the Japanese puppet state of Manchoukuo in Manchuria and Jeho was declared independent on February 18, 1932. This not only gave Japan the control of some of the best lines of the Chinese railways, but also effectively stopped all Chinese efforts at integrating the territory. An independent Manchoukuo would certainly have implications on the status of the Chinese Eastern Railway which Japan had long wanted to take over.[2] Through the new government, Japan also was better able to consolidate her hold of Manchuria and Mongolia in the face of Chinese and Russian competition, and to prepare for the ultimate conquest of the Chinese mainland, in accordance with the spirit expressed in Tanaka's memorial. To achieve these objectives, the Hsinking[3] Government, on the day of the first anniversary of Manchoukuo, announced a vast economic plan which envisaged the

---

[1]Huang Po-chiao, I ko wu-nien chien chih Ching-Hu, Hu-Hang Jung tieh-lu chung-wu shing-cheng [An administrator-general of the Peking-Shanghai and Shanghai-Hangchou-Ningpo railways for five years] (n.p., 1940), pp. 176-77. See also Ling Hung-hsun, Chung-kuo t'ieh-lu kai lun, p. 14.

[2]This can be evidenced by the constant pressures the South Manchurian Railway Company put to bear on the Chinese Eastern Railway in terms of freight war, the construction of lines that competed with the Chinese Eastern Railway and cut across its territory, and to standardise the gauge of the line between Harbin and Ch'angch'un.

[3]Even the choice of Hsinking (Ch'angch'un) as the new capital was partly based on its nodal position in terms of railway transportation. See Norton Ginsburg, "Ch'ang-ch'un," Economic Geography 23 (October 1947): 290-307. The nodality of Ch'angch'un, as is shown in this section, was greatly increased with the addition of many new railways after 1932.

expansion of the railway system to 25,000 kilometers and the construction of at least 4,000 kilometers in the following ten years.[1] Japan's position in Manchoukuo, particularly in northern Manchuria, was finally firmly established in 1935, when the Chinese Eastern Railway was sold to Manchoukuo and its gauge was immediately standardised.[2]

Japan's railway strategies in Manchuria during this period can be summarised in three aspects:

1. To construct more direct and shorter linkage routes with Korea so that the distance between Ch'angch'un (Hsinking), capital of Manchoukuo, and the Japanese homeland could be shortened, and that Talien could be bypassed if necessary. Thus, the Chilin-T'umen line connecting the Korean port Chongjin was quickly completed in 1932.[3] In addition, a third linking route was made possible by the completion of the Ssupingchieh-Meihok'ou line in 1936 and of the Meihok'ou-Chian line in 1937, just before the Japanese offensive started.
2. To extend Japanese controlled railway lines into northern Manchuria, not only to weaken the influence of the Chinese Eastern Railway, which until 1935 was still very much in the hands of the Russians, but also to strengthen Japan's hold of the area. Thus, reaching out from the Chilin-T'umen-Chongjin line, three extensions were constructed. To the north, the Lafa-Harbin lines was built in 1934, bypassing the Russian held Ch'angch'un-Harbin segment of the Chinese Eastern Railway and connecting with the Chinese built Harbin-Hailun line which was extended to Peian and ultimately to Heiho on the border in 1935. On the east, the construction of the entirely new line from T'umen through Mutanchiang to Hut'ou, again a border town, and branching out at Link'ou to Chiamussu, commenced in 1933. On the west, the Ch'angch'un-T'aoan line, finished in 1935 and connected with the Chinese-

---

[1] Japan Manchoukuo Year Book, 1934, p. 647. It also states, " . . . the Chinese Eastern Railway, which means more to Manchoukuo in the north than the South Manchurian Railway does in the south . . It is not only a trunk line for international traffic . . . but is the main artery of economic communications in North Manchuria."

[2] The deal was without reference to China, though still a partner of the line.

[3] It should be noted that the line had been completed by Chinese up to Tunhua, and its extension to the Korean border had long been strongly objected to by the Chinese before 1931.

built T'aonan-Paich'engtzu line on the western end and the then operative Ch'angch'un-T'umen-Chongjin line on the eastern end, provided through traffic between Chongjin and Inner Mongolia. It was designed to run parallel to the Chinese Eastern Railway and was to reach Aershan on the western border. By 1937, the line reached Solun, less than 200 kilometers from the border.

3. To construct railway lines into north China. Thus in 1933 construction was started on a line parallel to the Peking-Shenyang railway between Ch'aoyang and Chengte, from which a branch line ran from Yehpaishou to Chihfeng a year later, penetrating into Inner Mongolia.[1]

Railways and Wars

Railway Development and the Sino-Japanese War

The outbreak of the Sino-Japanese War in 1937 marked another turning point in the evolution of the railway system and of the Chinese space-polity. Sun Yat-sen's doctrine of san-min chu-i for an egalitarian, democratic, and prosperous society was overshadowed by practicalities of a life-and-death struggle; rising nationalism which had the overtone of self-strengthening aspired by the desire for national rebirth was more sharply focused by the war of resistance;[2] and the context in which various railway policies which had most of the time been directed at transforming the space-polity of China had to be completely and abruptly changed. Whereas railways had previously been generally designed to generate more and more traffic and were therefore constructed to anticipated capacities, war-time transportation and particularly railway planning had to be based on "minimum transportation requiarement in a war-time environment," measured against a "subjective maximum mobilization."[3]

It was not surprising that when war started, the railway system immediately became the most strategic element. Areas and cities along railway lines became battlefields one after another,

---

[1] Much of this section is derived from Ling Hung-hsun, Chung-kuo t'ieh-lu kai lun, pp. 16-17.

[2] The desire for national rebirth had given rise not only to the October 1911 Revolution and the establishment of the Republic of China, but also Communism in the 1920's. Nationalism was most sharply focused by the formation of the United Front between the KMT and CCP at the height of the anti-Japanese war in the late 1930's.

[3] Ch'en Hui, Kuanghsi chiao-tung wen-t'i [Communications problems in Kuanghsi] (Changsha: Commercial Press, 1938), p. 5.

and military successes and failures directly related to the control of important railway lines.[1] As the war rapidly involved the ecumene of China, that is North, East, and Central China, railway policies had a clear emphasis on developing and strengthening the rear and interior, and on international connections. Throughout this period when construction progressed in the midst of serious destruction, the following characteristics in railway development are worth noting: (1) the policies of contesting directly with the enemies in areas of conflict, or even of strengthening of the ecumene, gave way to those protecting the outer zone and periphery, namely, the Southwest and Northwest; (2) not only were no railways constructed in areas of war operation, but rolling stock and railway equipment were rapidly removed in retreat from the front for construction in the rear; (3) wartime construction, constrained by difficult terrain and the lack of materials, had a low standard; (4) ever since the outbreak of war, most railway construction was financed directly by the central government, instead of by foreign loans or profits from other lines.[2]

Thus, as soon as the war started, construction of four important lines in the Southwest was undertaken, namely, the Hsiang-Kuei (Hunan-Kuanghsi) line, the Tien-Mien (Yunnan-Burma) line, the Hsu-Kun (Ssuchuan-Yunnan) line and the Ch'ien-Kuei (Kueichou-Kuanghsi) line. The Hsiang-Kuei line, to run from Hengyang to the Vietnamese border at Chennankuan, was designed as an international supply line in anticipation of the loss and blockage of the newly completed Hankow-Canton line. This line reached Kueilin in 1938 and Liuchou a year later. Japanese landing from Peihai and the occupation of Nanning in November 1939 interrupted the completion of the line to Chennankuan through Hanning. By late 1938, a protracted war seemed inevitable, and the need to extend the Hsiang-Kuei line further into the interior was acutely felt. However, the construction of the Ch'ien-Kuei railway proceeded only slowly. It did not reach Ishan until 1940, Chinch'engchiang until 1941, and Tuyun until 1944. Rails, rolling stock, and other equipment for its construction were removed from the Che-K'an (Chechiang-Chianghsi), Hsiang-K'an (Hunan-Chianghsi), and Yueh-Han (Canton-Hankow) lines.[3] The

---

[1]Ling Hung-hsun, Chung-kuo t'ieh-lu kai lun, pp. 17-18.

[2]With the exceptions of the Nanning-Chennankuan section of the Hunan-Kuanghsi (Hsiang-Kuei) line and the Yunnan-Ssuch'uan (Kunming-Hsufu) line which were financed by French loans. However, these loans were not guaranteed with the properties of the lines as they usually were. See Ling Hung-hsun, Chung-kuo t'ieh-lu kai lun, pp. 21-22.

[3]I Yeh, Kuei-Chien lu shang tsa i [Memoirs on the Kuei-Chien railway] (Hong Kong: Chih Cheng Press, 1971), p. 104.

construction of the line progressed so slowly that it was not much ahead of cities falling into Japanese hands one after another, and the completion of each section therefore became so vitally important for military and civilian transportation.[1]

Again designed to form another international supply line, the Hsu-Kun and Tien-Mien lines would have run from Hsufu on a tributary of the upper Yangtzu River in Ssuch'uan through Kunming to Mengting on the Burmese border. The construction of the former depended on materials coming through the French-built Yunnan-Vietnam line which was blockaded in 1940 when Japan occupied Vietnam. The only section of 174 kilometers completed between Kunming and Chani was built with equipment removed from the Kunming-Vietnam line. Likewise, construction of the Tien-Mien line had hardly started before it was stopped by the Japanese occupation of Rangoon in 1941.[2]

Apart from these, the east-west trunk route, the Lung-Hai line, was extended beyond Hsian between 1939 to 1945, reaching T'ienshui shortly after the war was over, with equipment and rolling stock removed from sections on the same line further east. In addition, a branch line from Hsienyang to T'ungkuan for coal supply was completed earlier in 1940 (fig. 7).

Most of this construction, together with existing lines in China outside Manchuria, fell into Japanese hands as the War progressed. Within one year after the outbreak of the War, 9,120 kilometers or nearly three-fourths of the railways in North and Central China and Kuangtung Province were lost to the enemy. All of the major battles were fought along the railway lines, and by December 1944 when the Japanese occupied Hunan and Kuanghsi Provinces, the Nationalist Government was in control of no more than 1,209 kilometers of railway lines, consisting practically of only part of the Lung-Hai line west of T'ungkuan and the narrow-gauge sub-system in Yunnan Province (table 8).[3]

---

[1]Ibid., and Pi I-tsun, Chuan shih hsi-nan tien-ti chien [The retreat to the Southwest] (Hong Kong: Chi Shih Nien Tai, 1973).

[2]Ling Hung-hsun, Chung-kuo t'ieh-lu kai lun, pp. 18-21.

[3]Yu Fei-peng, Shih-wu-nien lai chi chiao-tung k'ai-fang Communications in the last fifteen years (n.p., 1946), p. 8; Ling Hung-hsun, Chung-kuo t'ieh-lu kai lun, p. 20; U.S. Army, Military Intelligence Division, Railroads of China 1-7 (Washington, D.C.: Strategic Engineering Study, SES 145, November 1944): 199, map no. 71-1c; and C. T. Chow, "China's Internal Railway Problems: The Case of the Railways' First Century, 1866-1966." (Ph.D. Dissertation, Michigan State University, 1972), p. 138, map 5.

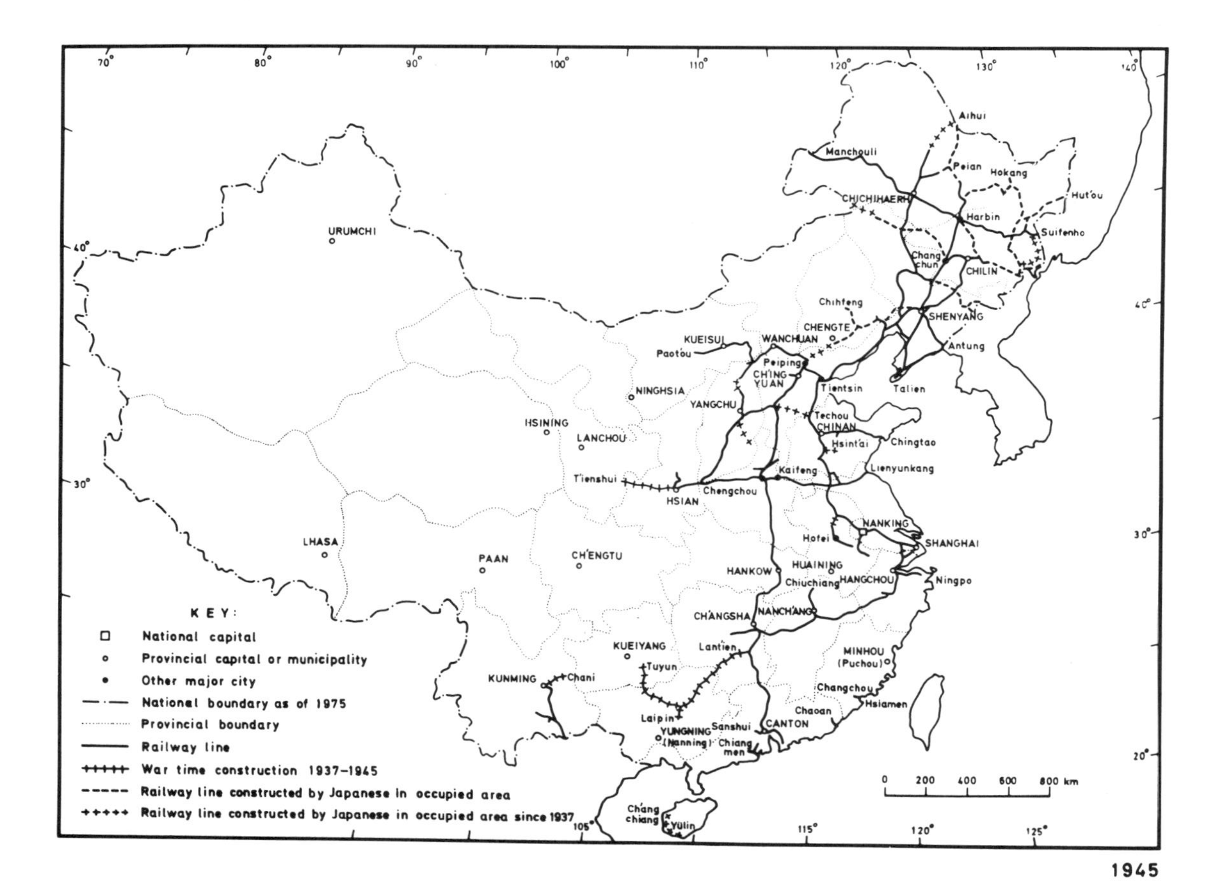

Fig. 7. Railways in China, 1945

For sources, see appendix D.

TABLE 8

CHANGES OF RAILWAY KILOMETRAGE DURING THE SINO-JAPANESE WAR (OUTSIDE MANCHURIA)

	from 18.9.1931[a] to 7.7.1937[b]	from 7.7.1937[b] to 12.1938[c]	from 12.1938[c] to 12.1944[d]	from 12.1944[d] to 8.1945[e]
Operating km at beginning of period	14,441	11,415	3,317	1,209
Constructed or repaired during period (km)	3,150 (new)	1,022 (new)	1,187 (new)	200 (repaired)
lost to enemy or removed (km)	6,176	9,120	3,295	–
Operating km at end of period	11,415	3,317	1,209	1,409

SOURCE: Simplified and adapted from Yu Fei-p'eng, Shih-wu-nien lai chiao-tung k'ai-feng, [Communications in the last fifteen years], (n.p., 1946), p. 8.

[a]The date of the incident that led to the establishment of Manchoukuo.

[b]The beginning of the Sino-Japanese War.

[c]The fall of Hankow (Hupei) and Canton (Kuangtung).

[d]The fall of Hunan and Kuanghsi.

[e]Japanese surrender.

Japanese Railway Strategies in Occupied China

The railway system in occupied China not only served as an important means of military penetration and control, but also played a significant part in the maintenance and supply of Japan's war economy. Thus, arteries fallen into Japanese hands were immediately utilized, and Japanese construction was specifically designed to facilitate shipments of troops and military supplies on the one hand, and to expedite the flows of resources and materials from the interior to the coastal ports on the other.[1] Keenly aware of the importance of an adequate railway network, the Japanese rehabilitated many lines previously built by the Chinese, converted others in part or whole from narrow to standard gauge, and constructed new rail links to enhance the strategic functions of some trunk routes.

Japanese railway strategies in occupied China can be discussed in relation to North, Central, and South China.

In North China, five new strategic links constructed by the Japanese were most significant: (1) the Peking-Jeho line completed the alternate route to the Peking-Shenyang railway for the movements of Japanese troops and for better integration between Manchoukuo and North China; (2) the Ningwu section from Yuanp'ing to Shouhsien of the Tungpu (Tatung to Puchou) line provided a north-south artery from the Peking-Suiyuan line to the Yellow River as well as the Lung-Hai line – also tapping rich coal and iron resources and facilitating troop movements; (3) the Hsinhsiang-K'aifeng cut-off, effectively joined the Peking-Hankow line with the Lung-Hai line, detouring the Yellow River break and Chinese guerrilla stronghold at Chengchou; (4) the Tungkuan-Luan line from Yangch'u (T'aiyuan) through southeastern Shanhsi Province was to hook ultimately into the Peking-Hankow and Lung-Hai railways and to relieve bottlenecks on the east-west Cheng-Tai line; and (5) the Shihchiachuang-Tehsien line, extending from the Cheng-Tai line and linking in with the Tientsin-Puk'ou line, in fact provided a straight route from Shanhsi to the port of Ch'ingtao, an alternate port exit to Tientsin. This construction gave Japanese occupation forces great manoeuvering flexibility and facilitated flow of needed materials.[2]

Unlike North China, railways in Central China were not combined into one vast network, but instead were linked for the most part to

---

[1] U.S. Bureau of Foreign and Domestic Commerce, Railway Transportation in Occupied China, 5 vols., (Washington, D.C.: The Bureau, Far Eastern Unit, 1943), pp. 28-35.

[2] Ibid.

shipping facilities along the Yangtzu River. Thus in this area where water routes provided both an alternate means of transport and a mode of junction links, new railway construction had been relatively insignificant.

On the other hand, in South China, owing to the wide-spread destruction of rail facilities by the retreating Chinese and the fact that Chinese forces still controlled certain sections of the trackage, the Japanese could utilize only limited portions of the railway system. Until the last two years of the war, both the Chinese and British sections of the Canton-Kowloon Railway and the Shant'ou rail system constituted the only lines controlled over their entire length by the Japanese. The Canton-Hankow trunk and the Canton-Sanshui branch lines, only partially controlled by the Japanese, were of limited strategic importance.[1]

Hainan Island, commanding the sea approaches to Eastern Asia and being an important stepping stone to Indochina, was important to the Japanese from both the military and economic points of view. Hence, shortly after the occupation of the coastal cities in 1939, the Japanese embarked upon a railway program to strengthen her hold of the island and to exploit its resources.[2]

Post-War Railway Plan and the Civil War

On the eve of the Japanese surrender in 1945, the Nationalist Government was in control of only 1409 kilometers of railway trunk lines in mainland China out of a total of probably some 25,120 kilometers including those on Taiwan and Hainan islands (table 8).[3]

At the same time, the Nationalist Government in Chungking, after consulting the Ministries of Communications, Defence, Economy, and Finance, for the first time in its history was able to work out a comprehensive five-year program for railway network development. This plan was explicitly based on the urgent political, economic, and transportational needs of the country. Politically, it was

---

[1]Ibid.

[2]Ibid.

[3]Yen Chung-ping, Chung-kuo chin-tai ching-chi-shih t'ung-chi tzu-liao hsuan-chi, tables 1 and 2, pp. 172-179. Note also that Yu Fei-p'eng, Shih-wu-nien lai chi chiao-tung k'ai-fang, gave a total of 30,205 km., whereas Ling Hung-hsun, Chung-kuo t'ieh-lu kai lun, Appendix, pp. 448-455, gave a total of 30,179 km.

aimed at connecting all provincial capitals by rail and the opening up of frontier areas in the Northwest and Southwest.[1] Economically, it adopted the principles of giving priority to areas of mining, heavy industry, and agricultural production. Transportationally, the objective was to link up different waterway systems, to complement road and river transportation, and to lessen railway congestion.[2] The plan envisaged the construction in five years of 13,886 kilometers of major trunk routes including the following: (1) the north-south trunk route in the interior from Kueiyang through Chungking, Ch'engtu, and T'ienshui to Lanchou, and from Lanchou extending northwestward to Hami, and northeastward to Paot'ou; (2) the extension of the Hunan-Kuanghsi line from Laipin to Lit'ang, and thence to Chanchiang and Chennankuan on the Sino-Vietnamese border; (3) the Hunan-Kueichou line from Hsiangt'an to Tuyun, and thence to Kunming through Kueiyang and Weining, completing the east-west trunk route in South China from Shanghai, and eventually continued to Mengting on the Sino-Burmese border; (4) the southward extension of the Nanking-Hsihsien line through Kueichi to Nanping, and thence to Minhou (Fuchou) and Changchou, forming the north-south trunk line with the Tientsin-Puk'ou line. The plan also included other minor lines such as the K'aifeng-Chinan line, the Sanshui-Liuchou line, the Ch'ingchiang-Ch'uchiang line, the Hankow-Hsiangyang line, the Changping-Shihlung line, and the Chihfeng-Tungliao line.[3]

However, this plan was never implemented due to the outbreak of the Civil War in the following year, in spite of the fact that most of the planned railways had been surveyed. Indeed, throughout the post-war years until the establishment of the People's Republic of China in 1949, not only that there was practically no new construction, but there was widespread destruction in China Proper as well as in Manchuria.[4]

---

[1]Chao Tseng-chueh, Chan-hou chiao-tung chien-she kai-lun [On post-war transportation reconstruction] (Shanghai: Commercial Press, 1947), pp. 35, 74 and 104; and Ling Hung-hsun, Chung-kuo t'ieh-lu kai lun, pp. 21-25.

[2]Chao Tseng-chueh, Chan-hou chiao-tung chien-she kai-lun, p.74.

[3]Ling Hung-hsun, Chung-kuo t'ieh-lu kai lun, pp. 21-25.

[4]Norton Ginsburg, "Manchurian Railway Development," pp. 398-411, stated that "railway construction in Manchuria had practically ceased after 1943," and footnoted that, "According to rough estimates made by . . . the Pauley Mission Soviet troops removed some 1,600 km. of railways."

Patterns and Functions of Railways in the Republic of China

During the Republican era from 1911 to 1948, some 13,760 kilometers of main lines and another 870 kilometers of branch lines had been constructed, bringing the network to a total of 22,780 kilometers of main lines and another 1,435 kilometers of branch lines in mainland China and Hainan Island (table 9). When it is realized that of the total mainline construction, the Japanese built 3,580 kilometers in Manchuria between 1931 and 1937 and another 874 kilometers mostly in other occupied areas after the outbreak of the Sino-Japanese War, the Nationalist Government's achievement compared less favorably with that of the Ch'ing Government. Indeed, over a long period of time both before and after the establishment of the Nanking Government in 1928 and until the September 1931 incident, the rate of railway development was much less than half of what had been achieved in the last 13 years of the Ch'ing dynasty. Moreover, if the Japanese construction in Manchuria is excluded, bringing the average yearly construction to only 493 kilometers for the period 1932-1937, not a single period in which the Republic's effort came anywhere near the peak of the Imperial Government.

No doubt, the Nationalist Government was seriously constrained by internal as well as external struggles, and the resultant network was as much a necessity of as a response to war-time conditions. The Japanese, on the other hand, must first have competed with Russia and China for the control of Manchuria and later expedited its development as a base for colonial expansion. Each of these exhibits clearly different patterns and functions.

The Patterns and Functions of Nationalist Railways

There was no doubt that railway development during the Republican period, like that under the Ch'ing Government, exhibited several distinct patterns, which, however, were not so much distinguished areally as was the case in the previous period. Rather they overlapped and transformed from one into another over time and space. Thus the semi-colonial pattern as a formal pattern with lines controlled by competing forces penetrating into the interior was gradually transformed. By 1931, the establishment of Manchoukuo in Manchuria unified the entire system and created a colonial pattern in the area. Within China Proper, the semi-colonial function of the railway network as a whole continued to underline railway development,

TABLE 9

RAILWAY CONSTRUCTION IN CHINA[a]
1912-1948 (in kilometers)

Period	Main line			Branch line		
	Total at end of period	Construction during period	average yearly construction	Total at end of period	Construction during period	average yearly construction
1911	9,016.01	—	—	561.75	—	—
1912-1927	12,619.62	3,603.61	225.23	866.35	304.60	19.04
1928-1931	13,634.62	1,015.00	253.75	1,010.42	144.07	36.02
1932-1937	20,173.10	6,538.48[b]	1,089.75	1,233.31	222.89	37.15
1938-1948	22,780.59	2,607.49[c]	237.05	1,435.29	201.98	18.36

SOURCE: Yen Chung-p'ing, Chung-kuo chin-tai ching-chi-shih t'ung-chi tzu-liao hsuan-chi, tables 1 and 3, pp. 172-179.

[a]Railway in Taiwan excluded.

[b]Including 3,579.75 km by Japanese in Manchuria.

[c]Including 874.50 km by Japanese in "Occupied China

while attempts by the Government of the Republic to achieve a nationalistic pattern centering at Nanking was repeatedly and seriously frustrated by construction which could only be rationalised as a war-time pattern.

With few exceptions, practically all of the trunk lines had been constructed with foreign interests,[1] whether in the form of direct control through joint ventures with China[2] or independent construction,[3] or in the form of indirect control through the offering of loans to China. In terms of railway loans alone, between 1914 and 1935, the contract amount increased by 44 per cent and the actual debts by more than 66 per cent, whereas the outstanding debts nearly doubled (table 10). These facts not only show that more of the contract loans were eventually taken than ever, but also that interest was accumulating at a faster rate than China's ability to repay. Over the same period, Japanese interests, of which no less than 70 per cent were in Manchuria, overtook those of Britain and jumped 8.7 times in terms of actual debts and nearly 12 times in terms of outstanding debts. The heavy indebtedness of the Chinese railways meant that as a system it was hardly able to generate much surplus. For decades, some of the lines struggled only to pay foreign debts. For example, from 1912 to 1935, the annual net incomes of some Chinese railways were much less than the amount of debt payment due (table 11).

The semi-colonial character of the Chinese railways was also evidenced by some of its functions. First, unlike other colonial systems, which were typically export-oriented, the system in China was better described as import-oriented. It is obvious that since the introduction of the railways toward the late Nineteenth Century, they greatly facilitated China's foreign trade. However, closer analysis reveals that as China's trade developed, her trade deficit widened, and balance was never really achieved until the end of the Republican period (fig. 8). It should be noted that the continuous

---

[1]With the exceptions within China Proper of the Tatung-Puchou line (1932-35), the Suchou-Chiahsing line (1935-36), and the Ch'ien-Kuei (Kueichou-Kuanghsi) line (1939-44), built by Central Government at various stages of the War, and several short lines such as the Chiangnan (Nanking-Hsihsien), the Huainan (Tienchiaan-Yuch'ik'ou), the Chaochou-Shant'ou, the Kochiu-Pisechai lines, etc., built by provincial companies.

[2]For example, the Chinese Eastern and the South Manchurian Railways.

[3]For example, the Antung-Shenyang line built by the Japanese military, and the Yunnan-Vietnam line leased to France.

TABLE 10

CHINA'S RAILWAY DEBTS BEFORE AND AFTER WORLD WAR I

Country	Contract Loan Amount					Actual Debt					Outstanding debt (including principal and interests)				
	1914		1935		1914	1914		1935		1914	1914		1935		1914
	Mill.¥	%	Mill.¥	%	=100	Mill.¥	%	Mill.¥	%	=100	Mill.¥	%	Mill.¥	%	=100
Japan	76.5	2.68	248.5	17.41	935.9	26.5	5.66	231.1	29.57	870.0	22.1	5.62	264.8	36.22	1198.7
Britain	405.5	40.95	436.6	30.57	107.7	194.0	41.29	214.0	27.43	110.5	177.0	45.04	144.9	19.83	81.9
Germany	78.3	7.90	78.4	5.49	100.2	75.9	16.17	76.1	9.73	100.2	74.4	18.93	100.4	13.74	135.0
France	234.7	23.71	257.5	18.03	109.7	77.9	16.58	100.3	12.83	128.8	77.3	19.67	89.3	12.21	115.4
Belgium	165.7	16.74	225.1	15.77	135.9	78.0	16.60	115.8	14.82	148.6	27.2	6.92	78.8	10.77	289.6
Holland	—	—	25.0	1.75	—	—	—	16.3	2.08	—	—	—	28.6	3.91	—
U.S.A.	79.0	7.89	106.3	7.45	124.5	17.0	3.62	26.6	3.40	156.4	15.0	3.82	23.6	3.23	157.5
Russia	0.4	0.04	50.4	3.53	1292.1	0.4	0.08	1.0	0.14	273.3	—	—	0.7	0.09	—
Total	990.1	100.00	1427.8	100.00	144.2	469.7	100.00	781.5	100.00	166.4	393.0	100.00	731.1	100.00	186.0

SOURCE: Ch'en Hui, _Chung-kuo t'ieh-lu wen-t'i_, p. 77, table III.

TABLE 11

PROPORTION OF RAILWAY DEBT PAYABLE TO NET INCOME, 1912-1935

Year	Shanghai-Hangchou-Ningpo Line			Nanking-Shanghai Line			Tientsin-Puk'ou Line		
	Net Income ('000 Yuan)	Foreign Debt Payable ('000 Yuan)	% of Net Income	Net Income ('000 Yuan)	Foreign Debt Payable ('000 Yuan)	% of Net Income	Net Income ('000 Yuan)	Foreign Debt Payable ('000 Yuan)	% of Net Income
1915	466	610	136.8	1,394	1,183	84.9	3,218	3,433	106.7
1920	412	605	146.8	2,692	637	23.7	8,508	2,092	24.6
1925	1,261	804	63.8	3,890	932	24.0	4,762	4,469	93.8
1930	1,715	1,410	82.2	3,786	3,448	91.1	2,176	6,633	304.8
1934	696	1,096	157.5	3,925	3,925	89.8	9,978	6,798	68.1

SOURCE: Yen Chung-p'ing, et. al., _Chung-kuo chin-tai ching-chi-shih t'ung-chi tzu-liao hsuan-chi_, table III.

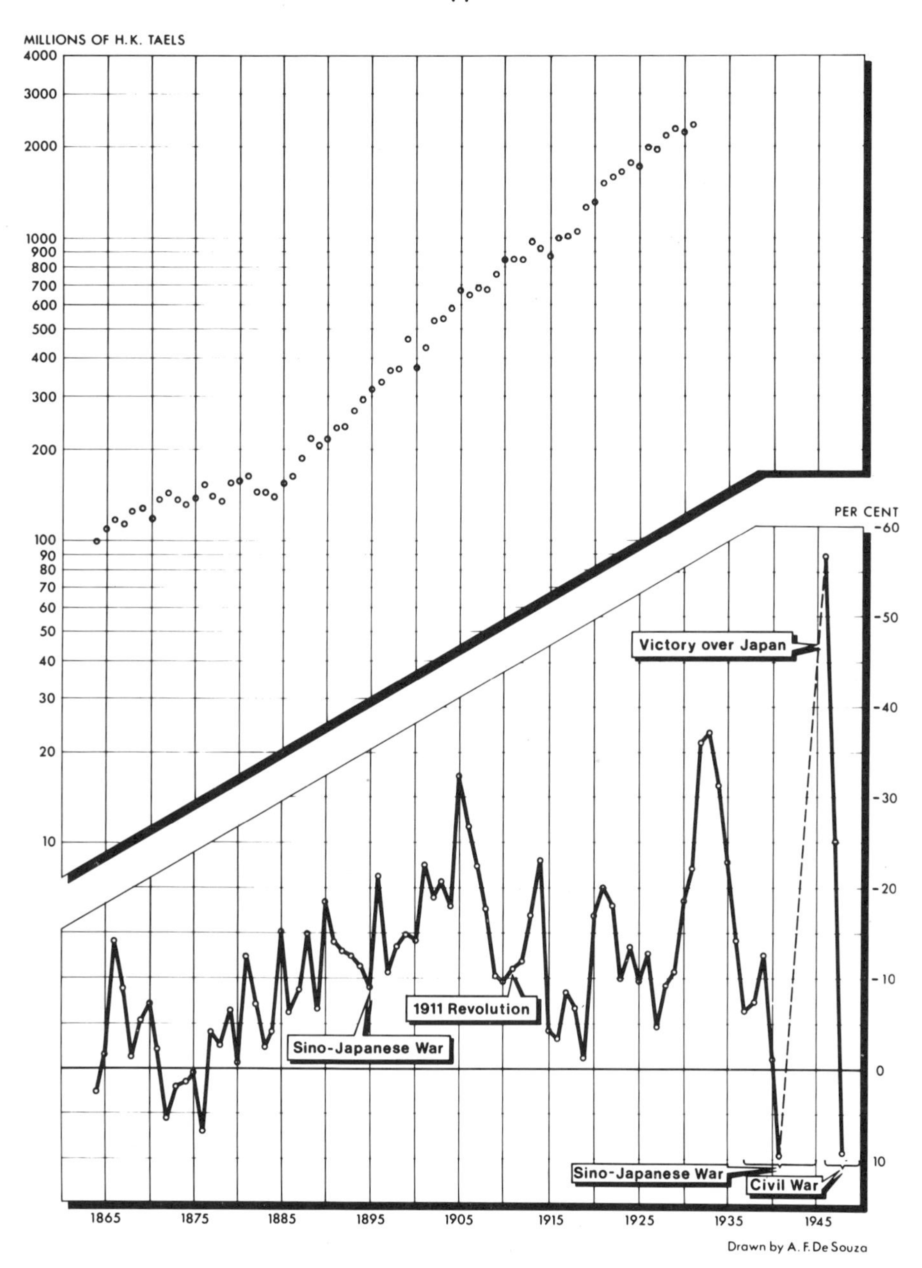

Fig. 8. China's Foreign Trade and Trade Deficit, 1865-1948

For sources, see appendix D.

trend of a widening trade deficit was only temporarily reversed with impending political crisis. Needless to say, imports in the beginning consisted mainly of cheap manufactured goods and consumables,[1] and had to rely on an efficient and cheap mode of transportation. The railway fulfilled this requirement admirably, since it was the cheapest per ton-kilometer freight mover in China[2] and, due to the strong influence of the foreign debtors, had a preferential fare structure which favored foreign goods and long distance shipment.[3] All these explained why no primate port city nor high-priority corridors were developed.[4]

Second, being the cheapest freight mover and for passenger traffic rivaled only by river junks,[5] the railway as a modern mode of transportation rapidly displaced other traditional modes.[6] As a factor facilitating the penetration into the interior and the expansion of markets for foreign goods, it contributed to the collapse of cottage industry and rural economy.[7]

Economically, Chinese railways throughout much of the Republican period had a high operating ratio. This ratio was nearly two-thirds higher than that of the South Manchurian Railway, and only the French leased Yunnan-Vietnam railway in Yunnan Province, which was maintained purely for political purpose, was even less profitable (table 12).

---

[1]Hsiao Liang-lin, China's Foreign Trade Statistics, 1864-1949, table 2, pp. 27-70. For example, cotton yarn, cotton goods, paper products, cigarettes, tobacco, etc.

[2]Arthur M. Shaw, "Transport Trends in China," Quarterly Review of Chinese Railways 1 (1 January 1937): 135-146, Table.

[3]Ch'en Hui, Chung-kuo t'ieh-lu wen-t'i, pp. 120-21, and P. L. Tan, "Traffic and Operation, Traffic Conditions of the Chinese National Railways," Quarterly Review of Chinese Railways 1 (1 July 1936): 57-68.

[4]See table 4, chapter II above.

[5]Arthur M. Shaw, "Transport Trends in China."

[6]By 1921, for example, more than 70% of goods arriving in Tientsin was transported by rail. See A. Rosenbaum, "China's First Railway."

[7]Fei Hsiao-tung, Hsiang-t'u chung-kuo [Earthbound China] (Hong Kong: Feng-huang Press, n.d., reprinted from 1947 version), pp.17-21.

TABLE 12

OPERATING RATIOS OF CHINESE, MANCHURIAN, AND FRENCH RAILWAYS IN CHINA, 1908-1940

	Chinese Railways (1917-1935)[a]	S. Manchurian Railways (1908-1940)[b]	Yunnan-Vietnam Railways (1916-1930)[c]
Mean (%)	61.37	37.60	75.72
Standard Deviation	8.42	5.30	10.57
S.D./Mean	0.14	0.14	0.14

[a]Calculated from Yen Chung-p'ing, et al., Chung-kuo chin-tai ching-chi-shih t'ung-chi tzu-liao hsuan-chi, p. 200, table 14.

[b]See table 16 below.

[c]Calculated from Ch'en Hui, Chung-kuo t'ieh-lu wen-t'i, p. 64, Table.

Indeed, the operating cost of the Chinese railways climbed steadily from the base year of 1917 to nearly 2.5 times by the late 1920s. Even after the Japanese annexation of Manchuria where many of the most expensive Nationalist railways laid (table 16 below), the operating ratio generally remained between 1.6 and 2.0 times that of the base year.[1] The high operating cost was attributable partly to the exceptionally high general expenses amounting to nearly a quarter of the total operating expenditure,[2] and partly to the fare structure. Since the preferential fare structure favored long distance shipments, average revenue per ton-kilometer rapidly decreased as average haul per freight ton increased.[3]

Another unique function of the Chinese railways was its direct contribution to political, particularly military, operations. Such military expediencies not only interfered with the building of a national system but also constituted the major features of a wartime situation. Ironically, railway loans provided the Nationalist Government an important means for raising funds. Between 1912 and

[1]Yen Chung-p'ing, et al., Chung-kuo chin-tai ching-chi-shih t'ung-chi tzu-liao hsuan-chi, table 12, p. 198.

[2]Ibid., table 13, p. 199.

[3]P. L. Tan, Traffic and Operation, pp. 57-68.

TABLE 13

OPERATING CONDITIONS OF CHINESE RAILWAYS, 1917-1935

Year	Revenue ('000 yuan)	Operating Cost[a] ('000 yuan)	Profit[b] ('000 yuan)	Outstanding Govt. Transport Payment ('000 yuan)	Outstanding Govt. Transport Payment as % of profit[c]
1917-18	63.87	32.54	31.33	5.96	19.02
1920	91.44	44.57	46.87	5.02	10.71
1925	127.52	77.64	49.88	17.99	36.07
1930	134.40	94.34	40.06	16.12	40.24
1935	171.09	113.01	58.08	12.88	22.18

SOURCE: Calculated from Yen Chung-p'ing, et al., Chung-kuo chin-tai ching-chi-shih t'ung-chi tzu-liao hsuan-chi, tables 14 and 15, pp. 200-201.

[a]Railway accounting in China at that time allowed depreciation only for engines and cars, thus actual operating costs should be higher than those shown in column.

[b]For the above reason, actual profit should be even lower than those shown in column.

[c]For the above reason, Government debts to the railway in relation to its profit should be even higher.

1927, for example, as much as the equivalent of 108.7 million yuan or 41 per cent of railway loans taken had been diverted mostly to military use by the Government.[1] In addition, each year Government debt to the railways for services rendered amounted to between 5 and 18 million yuan during the period from 1918 to 1935, or an equivalent of between 11 and 40 per cent of the entire system's book profit had it been paid up (table 13). In other words, the Chinese railways heavily subsidized Government political and military operations. Before 1931 and in connection with the Northern Expedition and the deteriorating situation in Manchuria, some of the lines such as the Peip'ing-Hankow and the Peip'ing-Shenyang lines performed vital functions for the transportation of troops and supplies.[2] During the Sino-Japanese War years, the remaining system in free China practically carried only freight and passengers directly or indirectly connected with war-time operations, i.e., troops, military supplies, and refugees, etc. Table 14 shows that the system carried more than 27 million troops and 5.4 million tons of military supplies during the war, while for the period between 1939 and 1945, when the Government of the Republic controlled only between 1,209 and 3,317 kilometers of tracks, merely 6 to 16 per cent of the total network (table 8), an average of more than 2 million troops and some 360,000 tons of military supplies annually passed through the grossly reduced system.

TABLE 14

MILITARY TRANSPORTATION DURING THE SINO-JAPANESE WAR

Period	Army (person)	Military Supply (ton)	No. of trains
July-Dec.			
1937	4,467,476	1,236,629	5,100
1938	6,985,360	1,633,161	5,859
1939	2,823,872	359,863	1,691
1940	2,915,725	475,984	—
1941	2,802,526	311,558	—

[1]Yen Chung-p'ing, et al., Chung-kuo chin-tai ching-chi-shih t'ung-chi tzu-liao hsuan-chi, table 7, pp. 190-193.

[2]Ibid., tables 23-24, p. 210.

TABLE 14-Continued

Period	Army (person)	Military Supply (ton)	No. of trains
1942	2,007,195	340,843	–
1943	2,984,456	455,849	–
1944	1,529,887	250,244	–
Jan-Aug 1945	916,556	365,124	–
Total	27,432,953	5,429,255	–

SOURCE: Kung Hsueh-sui, Chung-kuo chan-shih chiao-tung shih [A history of China's war-time communications] (Shanghai: Commercial Press, 1947), pp. 166-175.

The Patterns and Functions of a Colonial System

Although complicated by Chinese efforts to integrate the area before 1931, the net processes of network development in Manchuria closely resembled those of a colonial system, if the initial phase of scattered ports development is substituted by the development of ports and foreign exits.[1] Penetration and feeder development were most vigorously pressed ahead during the 1920s when China and Japan competed through railway construction to strengthen their hold of the area.[2] Interconnection was practically completed by 1937, when the increasingly dominating role of Talien and the major corridor formed by the Chinese Eastern and the South Manchurian Railways were only too apparent. The Manchuria system too was typically outward-oriented, unlike that in China Proper.

During the first two decades of the Republic, Japanese railway interest in Manchuria rapidly expanded through Sino-Japanese cooperation and through the offering of loans to China. By 1931, though Japan had direct or indirect interests in only some 38 per cent of the network, she had contributed nearly 50 per cent of the capital

[1]Edward J. Taaffe, et al., "Transport Expansion."

[2]T. A. Bisson, "Railway Rivalries in Manchuria between China and Japan," Foreign Policy Reports 8 (13 April 1932): 29-42.

invested in Manchurian railways (table 15). This is partially explained by the fact that Japanese loans cost 9 per cent or more per annum and were often used to finance road construction before the lines were in actual operation.[1] Consequently, lines constructed with Japanese loans not only had unusually high capital-output ratios

TABLE 15

OWNERSHIP AND CONTROL OF RAILWAYS IN MANCHURIA, 1930-1931

	Capital Invested		Railway Operating	
	Chinese Yuan (millions)	%	Km	%
Total	1000[b]	100.0	6300	100
Japanese Interests:				
direct investment	494[c]	49.4	1134[c]	18
Sino-Japanese			252[c]	4
loans			1008[c]	16
Russian Interests				
Sino-Russian[a]	349[c]	34.9[a]	1764[c]	28
Chinese Ownership	157[c]	15.7	2142[c]	34

SOURCE: Yuan Wen-chang, Tung-pei tieh-lu wen-ti, (1932), p. 96, and Chou Shun-hsin, "Railway Development and Economic Growth in Manchuria," China Quarterly 45 (January-March 1971): 57-84.

[a]This refers to Russian-controlled investments which included French and other foreign interests for the construction of the Chinese Eastern Railways, ownership of which was only nominally Sino-Russian.

[b]This total included investments in yuan, yen, and roubles. Conversion was based on the ratios: 1 yuan = 1 yen = 1 rouble. At the exchange rate of 2 yuan = 1 U.S. dollar. This was equivalent to 500 million US dollars. It excluded earnings reinvested and other surplus accumulations.

[c]Calculated from given percentages.

during the early stages of their operations, indicating heavy capital investment in relation to operating revenue, but also extremely

---

[1]Chou Shun-hsin, "Railway Development and Economic Growth in Manchuria," pp. 63-65.

high operating ratios yielding low profits (table 16). By comparison, the South Manchurian Railway enjoyed the advantage of having the lowest capital-output and operating ratios among the major lines in Manchuria and was therefore the most profitable line. Its profitability, both in terms of total earning and earning per kilometer, was four times larger than that of all the railways in Japan combined, and six times greater as compared with those in Korea.[1]

TABLE 16

CAPITAL-OUTPUT AND OPERATING RATIOS OF MAJOR MANCHURIAN RAILWAYS, 1903-40, SELECTED YEARS

	SMR	CER	PMR	KCR	STR
		Capital-Output Ratios, 1907-40			
Mean (%)	2.30	16.21[a]	3.94	4.28	5.69
Standard Deviation	0.40	—	0.94	1.46	2.47
S.D./Mean	0.17	—	0.24	0.34	0.43
		Operating Ratios, 1903-40			
Mean (%)	37.60	62.50	42.60	73.50	65.20
Standard Deviation	5.30	11.03	10.28	9.69	13.13
S.D./Mean	0.14	0.18	0.24	0.13	0.20

SOURCE: Chou Shun-hsin, "Railway Development and Economic Growth in Manchuria," tables V, XIV, and XV, pp. 68 and 79.

[a]Mean of two sets of estimates for the years 1903-1906 by J. Arnold and *The China Year Book*, 1931-1932.

Abbreviations: SMR = the South Manchurian Railway
CER = the Chinese Eastern Railway
PMR = the Peiping-Mukden (Shenyang) Railway
KCR = the Kirin (Chilin)-Ch'angch'un Railway
STR = the Ssup'ing-T'aonan Railway

Extensive Japanese influence in Manchurian railways thus enabled her not only to neutralise Chinese efforts to reintegrate the area through railway construction, but also to successfully compete against Russian influence in northern Manchuria, to the extent that the direction of the freight traffic of the Chinese

[1]*Japan Manchoukuo Year Book*, 1934, p. 648.

Eastern Railway was reversed and the monopolistic position of the South Manchurian Railway eventually was firmly established by the 1920s (table 17). Network development after 1931 further reduced the importance of the Russian-built line, while tightening Japanese control of the area.

TABLE 17

DIRECTIONS OF THE CHINESE EASTERN RAILWAY'S FREIGHT TRAFFIC, 1908-24, SELECTED YEARS

Year	To the Chita Railway	To the Ussuri Railway	To the SMR	From the Chita Railway	From the Ussuri Railway	From the SMR
	(Total outflow traffic via these three lines = 100%)			(Total inflow traffic via these three lines = 100%)		
1908	9.1	90.9	0	21.4	78.4	0
1910	5.3	83.5	11.2	6.3	40.0	53.7
1912	5.3	83.1	11.6	11.9	25.1	62.9
1914	3.7	79.9	16.3	10.9	23.2	65.9
1916	3.6	67.6	28.8	5.5	23.3	71.2
1918	1.5	25.7	73.0	0.6	16.5	82.9
1920	7.7	11.1	81.2	1.6	4.6	93.8
1922	5.6	39.3	55.1	2.1	5.6	92.3
1924	0.7	39.3	60.4	2.6	7.0	90.4

SOURCE: Chiao-tung shih lu-cheng pien, vol. 17, pp. 332-340.

Characteristically, the most important feature of the Manchurian railway system as a whole, whether when it was separately managed before 1931 or when it was centralised by the colonial empire of the South Manchurian Railway Company after that, was its export function. Between 1914 and 1933, from 64 to 93 per cent of Manchuria's annual pig iron production, from 44 to 58 per cent of her coal production, from 33 to 99 per cent of her soy-beans production, and from 35 to 81 per cent of her salt production were exported.[1] Since most of these exports relied heavily upon railway transportation, the importance of rail service to the export activities was obvious. Indeed, to take Suifenho and Talien, important

[1]Chou Shun-hsin, "Railway Development and Economic Growth in Manchuria," pp. 60 and 81.

terminal cities on the Chinese Eastern and the South Manchurian Railways respectively, as examples, export freight was consistently more than 15 times those of import at the former and from 2.26 to 4.23 times at the latter cities (table 18). This export function of the system is also reflected from the revenue structure which was clearly freight-biased. Specifically, income from freight typically contributed to two-thirds or more of the total income of all of the major Manchurian railways between 1903 and 1931.[1] Whereas before 1931 the freight traffic of the Chinese Eastern Railway consisted mainly of agricultural products, that of the South Manchurian Railway was made up largely of coal shipments. After the establishment of _Manchoukuo_, several changes in the pattern of freight traffic occurred: (1) the rapid expansion of shipments of minerals, including coal, accounting for as much as 55.3 per cent of the total freight traffic by 1935; (2) the emergence of manufactured goods which by the mid-1930s rivaled agricultural products as an important item of freight movements; and (3) the relative yet persistent decline of agricultural freight.[2]

TABLE 18

IMPORT-EXPORT RATIOS OF THE MANCHURIAN RAILWAYS
1925-1929

Year	via Suifenho	via Talien	Total
1925	16.60	3.74	5.19
1926	15.57	3.02	4.91
1927	18.17	2.80	5.14
1928	15.98	2.26	4.36
1929	17.85	4.23	5.44

SOURCE: Based on Ch'en Hui, _Chung-kuo t'ieh-lu wen-t'i_, p. 98.

The colonial function of the railway system in Manchuria under the South Manchurian Railway Company is further evidenced by an

---

[1]These include the South Manchurian Railway, the Chinese Eastern Railway, the Chilin-Ch'angch'un Railway, the Ssupingchieh-T'aonan Railway. See Chou Shun-hsin, "Railway Development and Economic Growth in Manchuria," table VI, pp. 70-71.

[2]Chou Shun-hsin, "Railway Development and Economic Growth in Manchuria," pp. 74-75.

examination of the Company's accounts. Table 19 shows that whereas the railways contributed between 82 and 91 per cent of the total known profit for the years 1931 and 1932, a huge amount of profit was diverted each year either to subsidize other colonial activities

TABLE 19

SOUTH MANCHURIAN RAILWAY COMPANY, PROFIT AND LOSS, 1931-1932

Business	Year ending March 31, 1931	Year ending March 31, 1932
Railways	58,562,154	48,185,482
Harbour	1,821,075	1,288,724
Coal Mines	1,813,172	16,938
Shale Oil Plant	32,568	289,669
Sundry Profits & Loss	1,535,447	1,310,890
Iron Works	666,633	-2,980,040
Local Public Works	-10,719,061	-10,877,411
Interests on Deposits and Loans	-18,506,991	-12,612,407
Overhead Charges	-10,867,790	-18,575,708
Depreciation Fund for Debentures	-1,330,480	-1,538,694
Hotels	...	-96,759
Interest Payable	...	-7,752,066
Retirement Allowance Fund Abolished	...	8,500,000
From Special Fund Abolished	...	7,500,000
.	.	.
.	.	.
.	.	.
Total Known Profit	64,431,049[a]	67,091,703[a]
Total Known Loss	-41,424,322[a]	-54,433,085[a]
Loss Omitted	-1,333,265[b]	-59,998[b]
Total Net Profit	21,673,462[c]	12,598,620[c]

SOURCE: Japan Manchoukuo Year Book, pp. 615-659.

[a]Obtained by adding all known profits or losses in the same year.

[b]Obtained by the formula: total known profit - total known loss - total net profit.

[c]Figure does not represent total of column given in original tables.

and expansion, such as the building of iron plants, public works, schools, hotels, etc., or to cover loan interest, overhead charges, and other interests that serviced the Company's subsidiary activities. As a result, the Company would have been in the red in 1932, had it not been able to make use of funds originally set aside for staff retirement allowances and other special purposes.

Summary of Findings

The 1911 Revolution and Sun Yat-sen's doctrine injected new elements into the *raison d'etre* of the Chinese state and provided some direction in China's rising expectation and aspirations. The disappearance of minority rule and the emphasis on nationalism of *san-min chu-i* removed some of the major internal contradictions that confused and constrained the long process of modernization and national integration, though widespread regionalism giving rise to warlordism continued to be a powerful centrifugal force. The important role of the railway in the unification of the peoples of China and in the integration of political territories was fully recognized. The uniform coverage of the Chinese political space by railway lines envisaged in Sun's national development plan was the best testimony of such a conviction.

However, this national goal, or political idea, of the Republic stood little chance of transforming the national political area through the construction of a modern transportation system. On the one hand, Sun's strongly nationalistic ideology, even when translated into a development plan, attracted no international capital, nor did it necessarily contribute to national unity in a multi-ethnic state such as that of China.[1] On the other hand, the almost simultaneous encroachment by advancing capitalism and maturing colonialism upon her extra-ecumenical and ecumenical areas rendered it impossible for China to systematically develop her transportation system, but instead forced her to respond passively and in a

---

[1]According to Moseley, "In her search for a political form, China has looked for a system of government which would take into account the special relationships existing between the Chinese core area and the non-Chinese parts of the empire . . . If the Ch'ing dynasty clung to an outmoded ideology . . . the situation of the Republic was worse. (there was) no workable intellectual system to draw upon, for the dominant ideology of the time was nationalism, a negative one from the stand point of frontier affairs . . . " See George Moseley, "The Frontier Regions in China's Recent International Politics," pp. 299-329.

piecemeal fashion. The attempt to re-integrate the Northeast hastened instead the annexation of the area by Japan which immediately turned it into a colonial system and a base for further colonial expansion. Within China Proper, the political idea was scaled down to an immediate policy of strengthening the capital by concentrating the limited resources for network construction around Nanking and in areas south of Yangtzu River. Even this policy suffered from prolonged military operations, as well as from a lack of geographical momentum, since the existing system had been centered at Peip'ing. That Nanking failed to become the center of an integrated network system and thus the nerve center of organization of the Chinese political space whether before or even at the end of the war, and the fact that Chungking, the war-time capital, never had any rail service at all, proved that the political idea never got very far.

Beyond the traditional Chinese ecumene, construction of lines into the Southwest, Central North, and Northwest was the result of war rather than national development priorities. To the extent that the post-war plan emphasized construction in these areas, it may be argued that the need to integrate the extra-ecumenical and even outer territories with the ecumene for the survival of the state was finally given high priority.

PART TWO

RAILWAY PATTERNS AND NATIONAL GOALS IN COMMUNIST CHINA

CHAPTER IV

RAILWAY TRANSPORTATION IN THE PEOPLE'S REPUBLIC OF CHINA

Objectives and Data

The railway system for the first time in the history of China was truly and completely centralised by the central government only since the establishment of the People's Republic of China. By and large, there were no conflicting ideologies or economic interests in the planning, development, and operation of the railway system after 1949. However, an exploration of the relationships between development objectives, which are largely governed by the prevailing political ideology, and the space-polity of Communist China is rendered difficult because of a serious lack of data. Indeed, in China other than the first two five-year plans, there has been practically no official public pronouncements on the systematic planning, construction, and functions of the transportation system, including that of the railway. Moreover, the Second Five-Year Plan was rather vague and indefinite. Elsewhere, public reports generally followed the Marxist attitude that transportation and communications are the vanguards of production and construction of national economy, and non-economic considerations appear to have been ignored.

In fact there has not been any comprehensive operational information either, though some very crude macro-data are available up to about 1960 concerning state investment, freight and passenger traffic, systems development, some indices of operating efficiency, etc. Unfortunately, such freight and passenger traffic data as are available up to 1960 are not directional and are not classified in any manner susceptible to analytical study. In any case, even such crude data have not been available since 1960.

One set of data, i.e., the development of the railway network at various stages, however, can be reconstructed from various sources. Such reconstruction produces a set of data not only consistent over time and space, but also consistent intra-systematically insofar as the evolution of the network in topological terms is concerned. Thus, in the absence of better information, this study of the railway patterns and national goals relating to the Communist period, while making use of any other data available, will begin with a careful reconstruction of the railway networks and a review of the place of the railway in the national transportation system.

Network Development

The Civil War that followed almost immediately the Sino-Japanese War not only interrupted the Nationalist Government's plan of railway development,[1] but also, like many previous struggles since the completion of the Wusung-Shanghai line, proceeded along major railway trunk routes. The control of the railway system was therefore vital to the fortune of the feuding parties. Table 20 shows that at the beginning of the Civil War, July 1946, the Government of the Republic was in control of more than 60 per cent of the network, and for the following two years this proportion remained above 50 per cent. The turning point came in June 1949 when the Government lost more than 80 per cent of the system and practically the War. Yet the Communists, even at this point, were in control of less than one-third of the national territory, some 59 per cent of the population, and slightly over half of the number of cities in China. Clearly the decisive factor in the Civil War was the railway system rather than territories, cities, or even population.

Table 20 also shows the extent of destruction to the railway system. Whereas 10,179 kilometers of railways were found in the liberated areas in July 1946, only 80 per cent of these were operative. Though more trackage was liberated in the following years, the proportion operative actually decreased to as low as 46 per cent in 1947 and was never restored even to the level of 1946 by the time the Government of the Republic fled to Taiwan in 1949.

The Rehabilitation of the Disintegrated System, 1949-1952

On winning the Civil War, the People's Republic of China inherited a railway system on mainland China and Hainan Island totalling at least 22,900 kilometers.[2] As a result of the Sino-Japanese War, Russian dismantling, and the Civil War, probably only about 17,000 kilometers were operative at some stage shortly before the

---

[1]Audrey Donnithorne, China's Economic System (New York: Frederick A. Praeger, 1967), p. 251.

[2]That is, total route length of 26,922 km (table 1) minus 3,960 km in Taiwan, as detailed in Ling Hung-hsun, Chung-kuo t'ieh-lu chih, pp. 448-455.

TABLE 20

RAILWAYS AND PEOPLE'S LIBERATION ARMY'S WAR PROGRESS
1946-1949

	July 1946	June 1947	June 1948	June 1949
Total land area (Km^2)	9,597,520	9,597,520	9,597,520	9,597,520
Area liberated (Km^2)	2,285,800	2,199,600	2,355,200	2,962,800
% of total	23.80	23.00	24.50	30.80
Total population ('000)	475,000	475,000	475,000	475,000
Population liberated ('000)	136,067	131,060	168,114	279,274
% of total	28.60	27.50	35.00	58.70
Total no. of cities	2,009	2,009	2,009	2,009
Cities liberated (No.)	464	417	579	1,061
% of total	23.00	20.70	28.80	52.80
Total length of railway (Km)	26,922	26,922	26,922	26,922
To be liberated (Km)	16,743	16,250	14,075	5,240
Kilometrage liberated (Km)	10,179	10,672	12,847	21,682
% of total	27.80	40.00	47.70	80.50
Km operative in liberated areas	8,200	4,951	8,132	17,185
% of liberated Km	80.56	46.39	63.30	79.26
% of total Km	30.46	18.39	30.21	63.83

SOURCE: Compiled from various tables in Chung-kuo jen-min chieh fang-chun tsung-pu, _Chung-kuo jen-min chieh-fang chan-cheng san-nien chan-chi_ [Tabulations of the Campaigns of the People's Liberation Army, 1946-1949] (n.p., 1949).

establishment of the People's Republic.[1] The system as a whole was therefore seriously broken up (fig. 9). The short line from Hsiaonanhai to Ch'ichiang, Ssuch'uan, which was constructed during the Sino-Japanese War, and the other from K'unming to Chani, Yunnan, which was part of the originally narrow-gauge French construction in the Southwest, had never been linked to the national network.

The first task of the new government after 1949 was undoubtedly the rehabilitation of the disintegrated system. Between early 1950 and 1952, i.e., the "period of economic recovery."[2] some 1,742 kilometers of pre-existing double or single track lines were repaired and revived, including the important north-south trunk route, the Hankow-Canton Railway, the Chengchiats'un to Fuyu line in the Northeast, and many others (fig. 10). In addition, the government began to modify some of the poorly constructed lines, such as the Paochi-Tienshui section of the Lung-Hai Railway,[3] and to build special purpose light railways for forestry and mining.

The new government was also active in building new lines. Construction of the Tienshui-Lanchou line, the last section of the East-West trunk route, the Lung-Hai Railway, was started as early as May 1950, long before the completion of the modification of the dangerous Paochi-Tienshui section. This line was to run through difficult terrain and underdeveloped areas where there were serious shortages of water, food, and fuel, let alone construction materials.[4] On October 1, 1952, the day when the Tienshui-Lanchou line was opened to traffic, construction of the more ambitious and more difficult line, the Lanchou-Hsinchiang Railway was started, since this line was to eventually link up with the Russian system beyond Hsinchiang and was to penetrate vast desert areas and earthquake belts.

In the Southwest, construction of several lines planned or unfinished by the Government of the Republic was undertaken. The

---

[1]Official reports since 1949 put figure at 11,000 km or half of the total system of 22,000 km operative.

[2]Tuan I-ming, Kung-fei chiao-tung chi yen-ch'iu [A study of communications in Communist China] (Taipei: Institute of National Defence, 1960).

[3]This section was constructed in 1943-45 with equipment removed from the Lung-Hai and Peiping-Hankow lines. See Chao Yung-hsin, Chung-Kung ti tieh-tao Kung-tso [Railway works in Communist China] (Hong Kong: Union Press, 1954), pp. 5-6; and Ling Hung-hsun, Chung-kuo t'ieh-lu chih, pp. 210-11.

[4]Chao Yung-hsin, Chung-kung ti tieh-tao kung-tso, pp. 7-8.

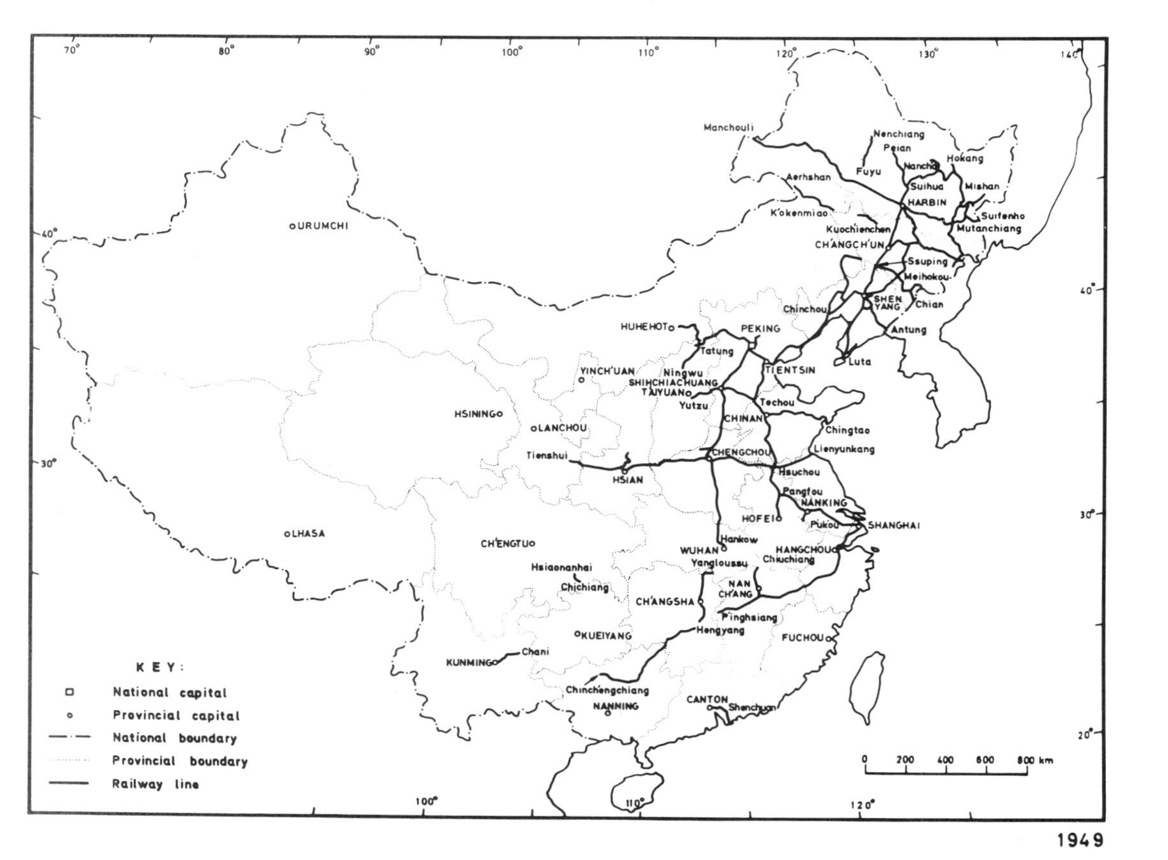

Fig. 9. Railways in China, 1949

For sources, see appendix B.

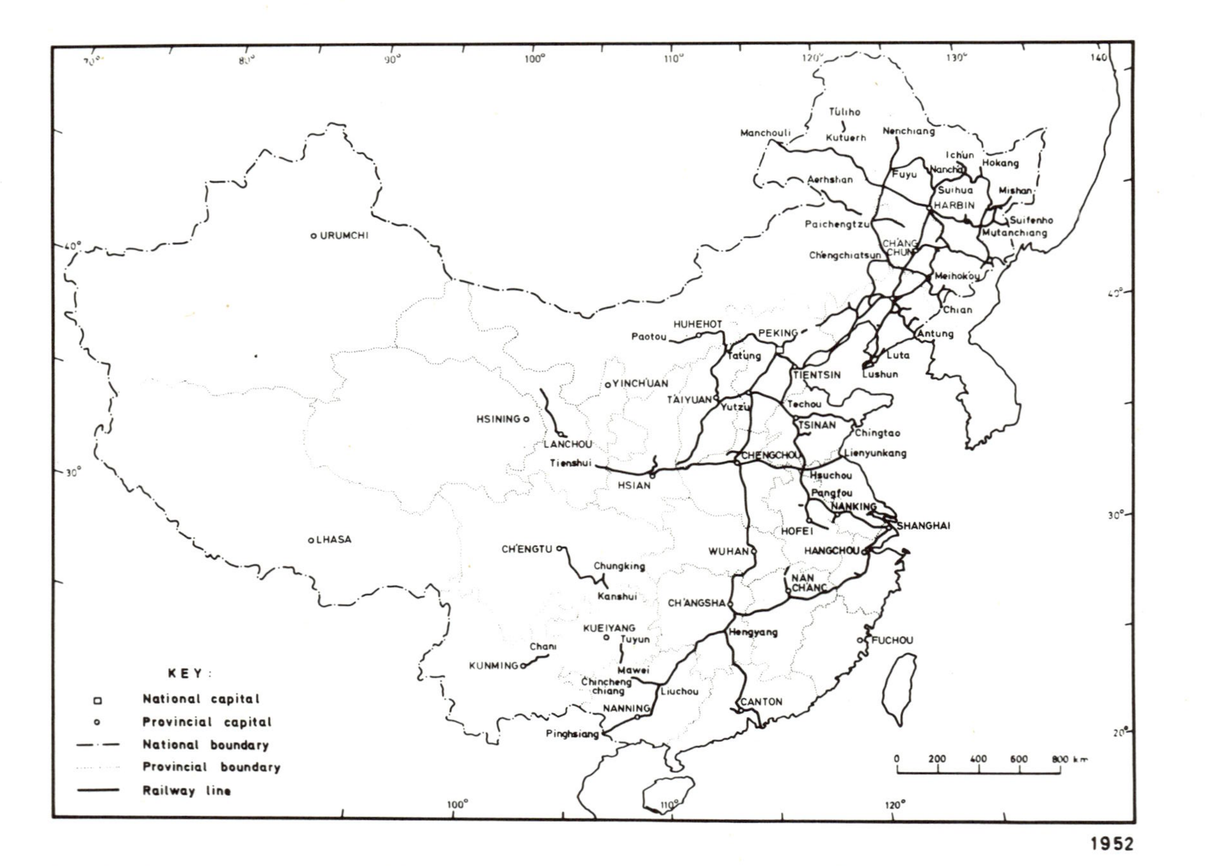

Fig. 10. Railways in China, 1952

For sources, see appendix B.

Ch'eng-Yu (Ch'engtu to Chungking) line, which had a long history of development and for which preparatory work was carried out until the end of the Civil War, was completed in 1952. The Lai-Mu (Laipin to Munankuan [now Pinghsiang] line, which was part of the unfinished Hsiang-Kuei (Hunan to Kuanghsi) line, was completed earlier on in February 1951 with equipment and materials obtained largely by converting the double track Canton-Sanshui line into a single track line.[1] Construction was also started on the Ch'ien-Kuei (Kueiyang, Kueichou to Liuchou, Kuanghsi) line, which had been completed up to Tuyun from Liuchou in the last years of the Sino-Japanese War but was seriously damaged in the Civil War period.

All together, new constructions added nearly 1,320 kilometers to the system, which at the end of the period had a total route length measuring 22,554 kilometers on the mainland and Hainan Island (appendix B).

The First Five-Year Plan, 1953-1957

The period of economic reconstruction commenced in 1953, when the government launched the First Five-Year Plan. As far as railway development was concerned, the Plan called for (1) the strengthening and modification of existing railroads, (2) the construction of some 4,084 kilometers of new trunk lines and another 3,284 kilometers of branch lines, and (3) the improvement of survey and design work.[2]

During the period, several of the new trunk routes planned were completed, providing better connections with the Soviet Union and the Mongolian People's Republic, such as the Chi-Erh (Chining to Erhlien) line (August 1955), or with a major trading port in South China such as the Li-Chan (Lit'ang to Chanchiang) line (July 1955), or improving the efficiency of the existing Peking-Paot'ou line such as the double track Feng-Sha (Fengt'ai to Shach'eng) line (July 1955) (fig. 11). The Lan-Yen (Lants'un to Yent'ai) line linking up Ch'ingtao with Yent'ai was completed ahead of schedule and was open to traffic in January 1956. Similarly, the Yingt'an-

---

[1] Ibid., p. 15.

[2] The Plan was approved retrospectively by the Party and the government in 1955, see Chung-hua-jen-min-kung-ho-kuo fa-chan kuo-min ching-chi ti ti-i-ko wu-nien chi-hua [The First Five-Year Plan of the People's Republic of China] (Peking, 1955). See also Jen-min shou ts'e [People's handbook] (Tientsin and Shanghai: Ta kung pao she, 1957), p. 526.

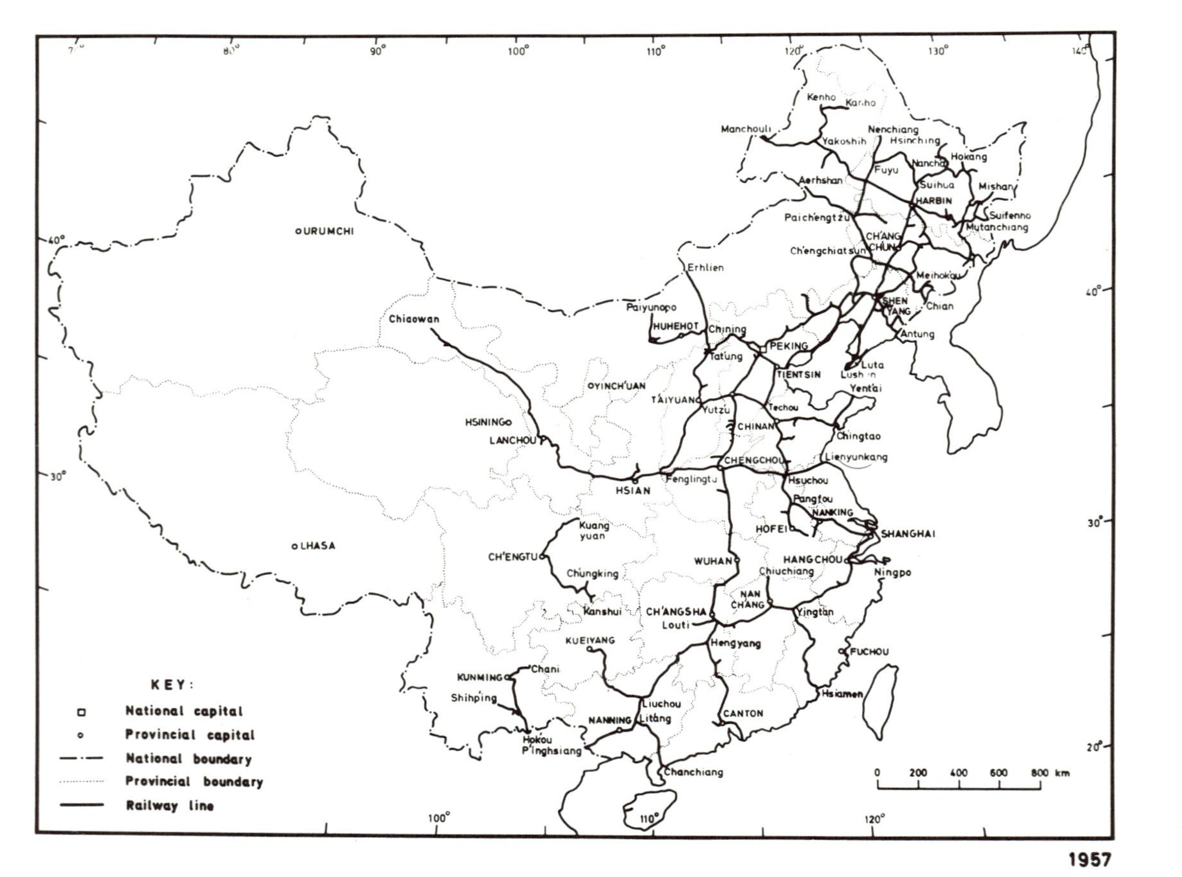

Fig. 11. Railways in China, 1957

For sources, see appendix B.

Hsiamen line, construction of which was planned only up to Yungan, was practically completed to Hsiamen by the end of 1956.

At the same time, construction work continued on the ambitious Lan-Hsin line, the Ch'uan-Ch'ien (Chungking to Kueiyang) line, the Pao-Lan (Paot'ou to Lanchou) line, while the Pao-Ch'eng (Paochi to Ch'engtu) line, for which track-laying was completed, apparently had serious engineering problems.

At the close of the period, 4,861 kilometers of new trunk lines had been constructed, nearly one-fifth more than planned, while double-tracking and repair each added roughly 900 kilometers. In terms of route length of major trunk lines, the system in 1957 measured 28,993 kilometers (appendix B).

The Second Five-Year Plan, 1958-1962

It is much more difficult to know how the government conceived railway development for the planned period, since the Second Five-Year Plan, which was passed one year before the end of the First Five-Year Plan in 1956, did not spell out details of any economic program.[1]

Major constructions during this period included the completion of lines either planned or actually under construction during the First Five-Year Plan period, namely, the Pao-Ch'eng line (January 1958), the Pao-Lan line (June 1958) and the Lan-Hsin line (January 1961). The Lan-Hsin line, also known as "friendship railway," was to connect with the Russian system beyond Hsinchiang through Karaganda but stopped short at reaching Urumchi (fig. 12). By comparison, other lines completed in the period were less spectacular, for example, the Nan-Fu (Nanp'ing to Fuchou) line (November 1958), the Lan-Ch'ing (Lanchou to Hsining) line (October 1959). Extensions from existing trunk routes were seen from K'unming to Chani (May 1960) and Ip'inglong (April 1959), from Wuhan to Tayeh and Suihsien (July 1958), and many others.

In short, probably just over 4,000 kilometers of new line were constructed, whereas repair work brought back into operation another 1,000 kilometers to the system, which at the end of the period had a total route length of approximately 34,091 kilometers (appendix B).

---

[1]See Chou En-lai, _Chung-kuo-kung-chan-tang ti-pa-ch'i ch'uan-kuo tai-piao ta-hui . . . ti ti-erh-ko wu-nien chi-hua . . ._ [The Second Five-Year Plan of the People's Republic of China . . .] (Peking, 1956).

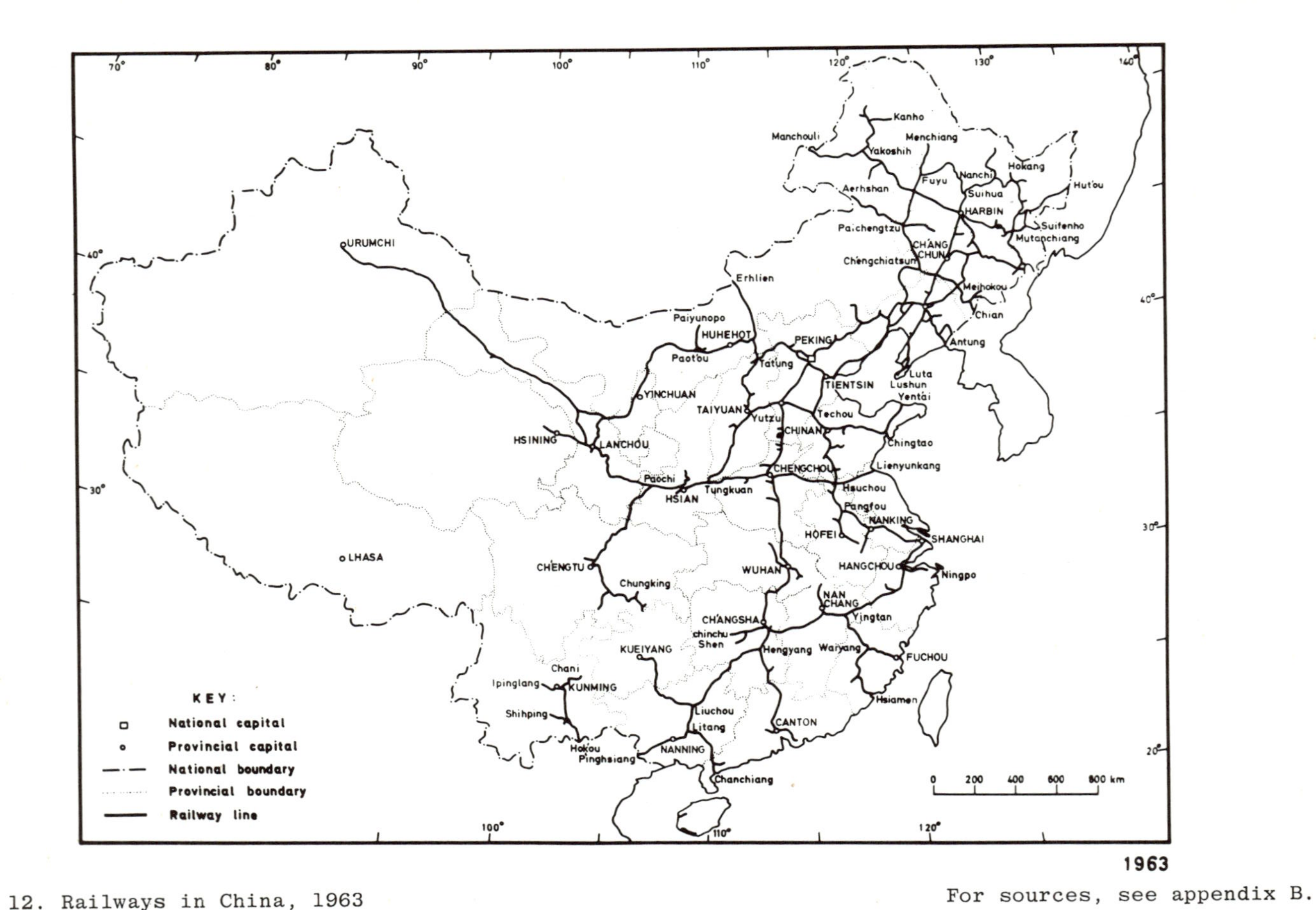

Fig. 12. Railways in China, 1963

For sources, see appendix B.

Of the new constructions, at least 1,332 kilometers or exactly one-third had in fact been completed in 1958 (table 21).

TABLE 21

RAILWAY CONSTRUCTION, 1950-1962

	Main and branch lines				
	Total	New	Repaired	Newly double tracked	Double track repaired
1950	808	97	427	–	284
1951	1021	743	138	–	140
1952	1233	480	605	–	148
Sub-total 1950-52	3062	1320	1170	–	572
1953	706	587	–	14	105
1954	1132	831	–	49	252
1955	1406	1222	39	87	58
1956	2242	1747	285	206	4
1957	1166	474	150	538	4
Sub-total 1953-57	6652	4861	474	894	423
1958	2376	1332	105	939	–
. . .	. . .	. . .	. . .	. . .	. . .
(1962)					
Sub-total 1958-62	8525	4027[a]	1071[a]	2174[b]	1253[b]

SOURCE: All except the last row from Wei-ta ti shih nien - Chung-hua jen-min kung-ho-kuo ching-chi ho wen-hua chien-she ch'eng-chiu ti t'ung-chi [Great ten years - statistics of the economic and cultural achievements of the People's Republic of China], State Statistical Bureau, ed., (Peking: People's Publishing Company, 1959), p. 60.

[a]Appendix B and Wu Yuan-li, The Spatial Economy of Communist China, pp. 252-261.

[b]Estimated based on Wu Yuan-li, The Spatial Economy of Communist China, table E-5, p. 271, extensively modified; Ralph W. Huenemann and Nicholas H. Ludlow, "China's Railroads," The China Business Review (March-April 1977), pp. 26-43; Ta Kung Pao, July 24, 1976; and Communist China Railroad Passenger Timetable, July-November 1963, JPRS no. 21963, (U.S. Department of Commerce, Washington D.C., 1963).

Throughout the period of the First and Second Five-Year Plans, 1953-1962, parts of several trunk routes were double-tracked, for example, the Peking-Wuhan line, the Wuhan-Canton line, the Lung-Hai line, and the Tientsin-Shanghai line. However, because of the interruptions of the Great Leap Forward and the Sino-Soviet split, apparently only the Peking-Wuhan line was fully double-tracked. Little is known indeed precisely when these double-track lines were constructed and completed. Thus, the sub-total figures for columns 4 and 5 in table 21 should be regarded as at best tentative.

The Cultural Revolution and the Third Five-Year Plan, 1966-1970

There was a complete dearth of information concerning railway development since the Second Five-Year Plan. In fact, there is now sufficient evidence to show that construction had been seriously slowed down, if not practically stopped, towards the end of the planned period. This was probably due to successive economic crises caused by exceptionally severe weather from 1959 to 1961; by the withdrawal of Russian experts in 1960; and by the philosophical debate over methods and priority of development which continued into the Proletariat Cultural Revolution (1965-68). There was not even a five year plan to follow immediately the Second Plan from 1963 onwards for some time. Before the start of the Cultural Revolution, the only railway of significance that had been built was probably the Ch'uan-Ch'ien (Chungking to Kueiyang) line (1965), of which the section between Chungking and Kanshui had long been built during the economic recovery period.

Steady construction seemed to have been revived only during the Third Five-Year Plan period, 1966-1970, particularly towards the latter part of the Cultural Revolution. Several important lines, the completion of which was not announced until the 1970's, were in fact constructed during the period, for example, the Tien-Ch'ien (K'unming to Kueiyang) line (1966), the Han-Tan (Wuhan to Tanchiangk'ou) line (1966), the Ch'eng-K'un (Ch'engtu to K'unming) line (1970), and the Chiao-Chi (Chiaotso to Chich'eng) line (1970) (fig. 13). Other identified lines included the important Nen-Lin (Nenchiang to Changling[1]) line in Heilungchiang.

By the end of the Third Five-Year Plan, all railway lines on

---

[1]The Nen-Lin line from Nenchiang through Linhsin to Moho on the border was completed up to Changling by 1970. See Ta Kung Pao, June 8, 1976.

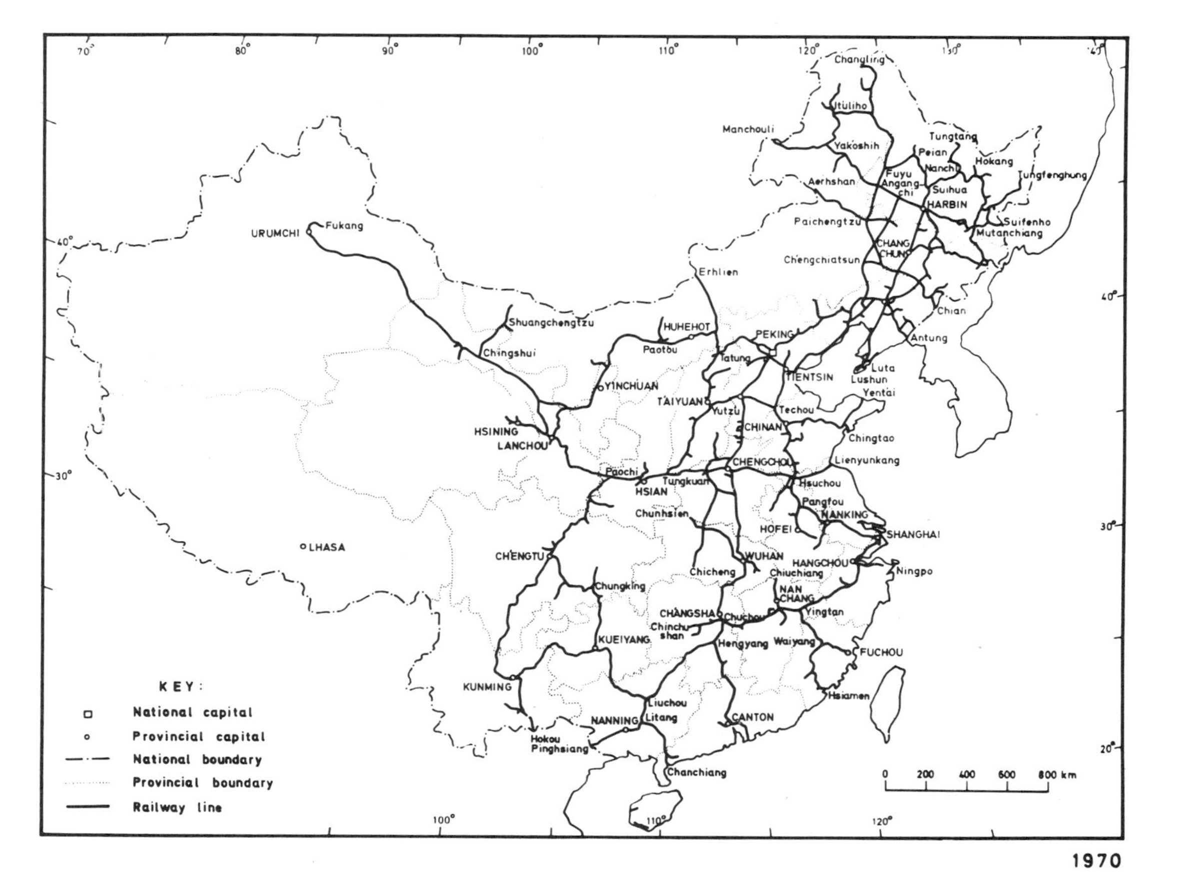

Fig. 13. Railways in China, 1970

For sources, see appendix B.

mainland China were for the first time in history integrated into a single system, and all provinces and autonomous regions, except Tibet, had access to the system, which had a total identified route length of 40,725 kilometers (appendix B).

The Fourth Five-Year Plan, 1971-1975

During the period of the Fourth Five-Year plan, construction of the Hsiang-Ch'ien (Chuchou to Kueiting) line (1972) completed the important east-west trunk route in South China from Hangchou to K'unming, paralleling the Lung-Hai line to the north; the Hsiang-Yu (Hsiangfan, Hupei via Ankang, Shenhsi to Chungking, Ssuch'uan) line, which provided Ssuch'uan better link with Central China, was completed in October 1973 but suffered serious damage from rainstorms and was not open to traffic until June 1, 1978 (fig. 14).[1]

Improvements to the accessibility of the national capital were provided by the construction of the T'ung-Ku (T'unghsien, Peking to Kuyeh) line (1975) and the central segment of the Peking-Yuanp'ing line. The period also saw in July 1975 the eventual electrification of one of China's most difficult lines, the 676 kilometer long Pao-Ch'eng railway,[2] and the construction of China's second electric railway, the Yang-An (Yangp'ingkuan to Ank'ang) line connecting the Pao-Ch'eng line with the Hsiang-Yu line.[3] Numerous extensions from major trunk routes, some penetrating into the border and others into mining and lumbering areas in the Northeast, North, Central, and Southwest China, were completed in the period. Whereas double-tracking of the Tientsin-Shanghai railway was yet to be finished, the period witnessed major decisions in electrification, especially of the lines over difficult terrain.

---

[1]At the time of topological reconstruction and analysis, no official announcement about the completion or even the construction of the Hsiang-Yu and Yang-An lines had been made. Their possible existence was first referred to in, Congress of the United States, China: A Reassessment of the Economy (Washington DC: U.S. Government, 1975), p. 267; Ta Kung Pao (Hong Kong), October 1 and November 18, 1977. Official announcements were subsequently made. See Ta Kung Pao, June 13, 1978.

[2]According to Ralph W. Huenemann and Nicholas H. Ludlow, "China's Railroads," pp. 26-43, the Peking-Shihchiachuang-T'aiyuan Railway is also an electric line, but this cannot be verified.

[3]Li Chien-ch'ao, "Wo-kuo yu i-t'iao tien-chi-hua tieh-lu - Yang-An tieh-lu" [Another electric railroad of our country - Yang-An Railway], Ti-li chih-shih, no. 7, 1978, pp. 1-3.

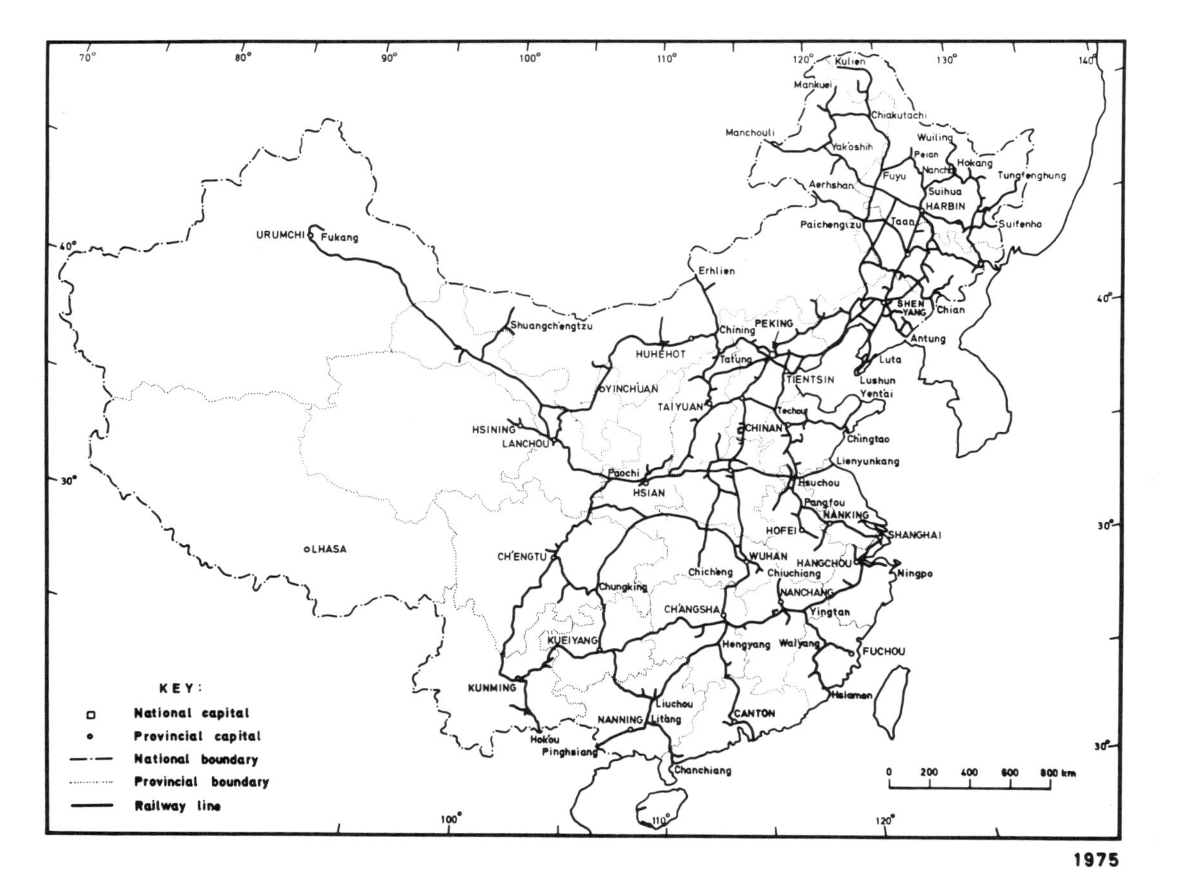

Fig. 14. Railways in China, 1975

For sources, see appendix B.

The Railway System and National Transportation

State Investment

The central government was responsible for providing investment funds in transport and communications. State investment in transport included the national railway system, but generally excluded local railways; national highways; civil aviation; and ocean, coastal, and river shipping services. Available statistics show (table 22) that transport and communications investment between 1952 and 1960 has been the second largest sectoral allocation, following investment in industry, accounting for between 13 and 21 per cent of total state investment. Investment in railways alone, however, consistently took up approximately two-thirds of investment in transport and communications.

Throughout the period from 1953 to 1960, i.e., the First Five-Year and the first three years of the Second Five-Year Plan periods, there was a steady increase in absolute terms in the investment in transport, except for 1957, a year of inflationary pressure and shortages of raw materials.[1] However, in terms of percentage of total state investment, it fell generally between 1955 and 1958, but rose rather sharply thereafter.

Table 22 also shows that whereas budgeted profits from transport and communications rose much less steeply than central government investment in the same sector, notably between 1956 and 1958, more than three-fourths of the budgeted investments could be recovered from revenues from the railway system alone during the First Five-Year Plan period. Over the same five years 1953-57 the railways, while absorbing 5,920 million yuan of state investment funds, returned to the state 6,900 million yuan in the form of surplus revenue.

Imports of Railway Equipment

The importance to the state of the national transportation system in general and that of the railway in particular can also be gauged by the values of China's import of transportation equipment. Between 1952 and 1973, there has been a steady but rapid increase in the import of transportation facilities and auxiliary equipment both

---

[1] Audrey Donnithorne, China's Economic System, p. 271.

TABLE 22

STATE INVESTMENT IN TRANSPORT AND COMMUNICATIONS, 1953-1960

	State Investment in Transport, Posts and Telecommunications		State Investment in Railways only		State Revenue (i.e. profits)	
	amount (mill. ¥)	as % of total State Investment	amount (mill. ¥)	as % of total State Investment	from Transport Postal and Telecommunications enterprises (mill. ¥)	from Railways alone (mill. ¥)
1952	760[a]	17	510[a]	12		
1953	1,070[a]	13	650[a]	8	)	
1954	1,500[a]	17	950[a]	10	)	
1955	1,760[a]	19	1,220[a]	13	)	6,900[g]
1956	2,610[a]	18	1,760[a]	12	2,132[d])	
1957	2,070[a]	15	1,340[a]	10	2,265[e])	
1958	3,400[a]	13	2,030[a]	8	2,388[f]	
1959	4,950[b]	19	3,550[b]	13		
1960	6,810[c]	21	5,000[c]	15		

[a] *Ten Great Years* (Peking: Foreign Language Press, 1960), pp. 57-59.

[b] Li Fu-ch'un, "Report on the Draft National Economic Plan for 1959," and Li Hsien-nien, "Report on the Actual Results of the 1958 State Budget and the Plan for 1959 State Budget," both in *Hsin-hua pan-yueh-k'an* [New China Semimonthly] (Peking), no. 9, 1959.

[c] Li Fu-ch'un, "Report on the Draft of the National Economic Plan for 1960," and Li Hsien-nien, "Report on the Actual Results of the 1959 State Budget and the Plan for the 1960 State Budget," in *Jen-min jih-pao* (Peking), March 31 and April 1, 1960, respectively; and Li Hsien-nien, "Report on Finance to the National People's Congress," *People's Handbook* (1960), p. 185. The 1960 figures specifically exclude "self-provided" investment *funds raised by local* authorities and enterprises, i.e., they exclude extra-budgetary funds. See also *Peking Review*, 5 April 1960, p. 11.

[d] *Ren-min Shou-ts'e* [People's Handbook], 1951, p. 214.

[e] Ibid., p. 218 (budgeted).

[f] Ibid., 1959, p. 229 (budgeted).

[g] For the First Five-Year Plan Period. See K. Pavlov, "China's Railroads," *Bulletin* 4 (September 1963): 12-28.

in absolute value and in relation to total imports of machinery, equipment, and transportation facilities (table 23). Specifically, it increased from a mere 1.39 million U.S. dollars or 7.9 per cent

TABLE 23

CHINA'S IMPORTS OF MACHINERY AND TRANSPORTATION EQUIPMENT: THOUSAND US$

	1952	1957	1962	1970	1973
Total machinery equipment and transportation facilities	174,878	357,279	57,598	291,672	631,329
Total transportation facilities and auxiliary equipment	1,392	20,931	19,499	129,347	354,057
(a) from Communist countries	1,215	13,748	16,739	18,896	83,657
(b) from non-Comm. countries	177	7,183	2,760	110,451	270,400
of which					
(i) Railway rolling stock and equipment	17	446	32	19,896	41,465
(ii) Motor vehicles	1,260	14,048	12,829	69,744	77,274
(iii) Motorless road vehicles	56	1,387	0	610	Ngel.
(iv) Vessels and port equipment	0	3,684	4,661	14,720	79,900
(v) Air transport facilities	1	0	1,971	23,426	115,219

SOURCE: People's Republic of China: Foreign Trade in Machinery and Transportation Equipment since 1952, U.S., C.I.A. Research Aid, A(ER) 75-60 (Washington, D.C., January 1975), various tables. Note that all statistics represent reported trade figures, and that totals do not necessarily equal the sums of relevant columns or cells.

of machinery imports in 1952 to over 44 million U.S. dollars or 56 per cent of machinery imports in 1973. Even in the year of heavy cuts in imports, namely, 1962, when five-sixths of China's machinery imports were cut, that of transportation facilities and auxiliary equipment entertained only a 6.8 per cent reduction in absolute value, and in fact it had achieved a jump in its share of total machinery imports from 5.8 per cent of 1957 to 33.85 per cent of 1962.

However, the railway system fared quite differently. In 1952 imports of railway rolling stock and equipment accounted for about 12 per cent of total transportation equipment imports, and it dwindled to 2 per cent in 1957 and a mere 0.16 per cent by 1962.

It was not until 1970 that it stabilised to a level of between 11 and 15 per cent of total transportation facilities imports. It should also be noted that whereas before 1957 China's import of railway rolling stock and equipment was almost solely from the Soviet Union, by 1962 Poland was the only Communist supplier country, and by 1970 it was Czechoslovakia which shared just about equally with Non-Communist countries China's imports of railway equipment. In 1973, although Czechoslovakia was still the only Communist supplier country, it accounted for a mere quarter of China's imports of railway equipment.

Freight and Passenger Traffic

No doubt, the railway plays an important role in the national transportation system of China, both in terms of freight and passenger traffic. Figure 15 shows that in 1949 the railway practically monopolised all freight movement, accounting for 56 million metric tons or over 83 per cent of all freight traffic. Although the relative importance of the railway in terms of freight originated declined over the years, due to the rapid development of other forms of modern transportation, particularly the highway system and to some extent inland and coastal shipping, the railway has since 1962 retained its dominant position, carrying some 60 per cent of all freight in China. Moreover, the strong position of the railway has hardly changed if measured in terms of freight turnover. Specifically, from 1949 to 1957 it was responsible for between 78 and 84 per cent of the total freight turnover in China, and probably for no less than three-quarters up to the present day (table 24).

Whereas railway freight volume has increased from 56 million metric tons of 1949 to 945 million metric tons of 1975, or 16.8 times, the per capita railway freight volume increased only from 0.103 metric tons to 0.951 metric tons, or 9 times over the same period. Similarly, whilst railway freight turnover of 458 billion metric ton-kilometers of 1975 represented nearly 25 times that of 1949, the average haul of railway freight increased by only 49 per cent from 1949 to 1957, and thereafter it stabilised at about 485 ton-kilometers.

For passenger transport, the railway is equally the dominant system, carrying no less than 530 million passengers in 1960. Although its share in total passenger trips decreased from 76 per cent in 1949 to 47 per cent by 1958 due to competition from road and

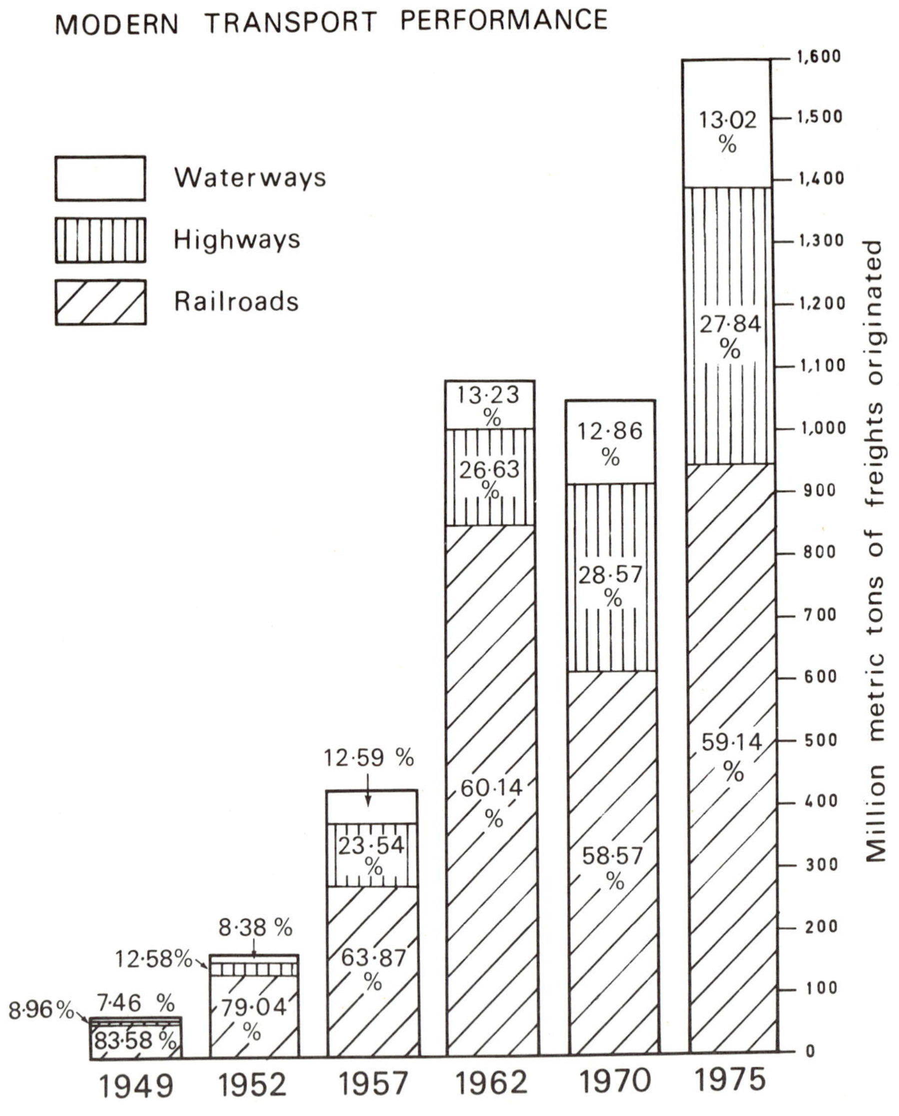

Fig. 15. Modern Transport Performance For sources, see appendix D.

TABLE 24

RAILWAY FREIGHT TRANSPORT PERFORMANCE, 1949-1975

	Total MMT	Freight originated Railway only			Total bil.MT/km	Freight Turnover Railway only		
		MMT	%	MT p.cap.[c]		bil.MT/km	%	ave. Ton/km
1949	67[a]	55.9[a]	83.6	0.103	23.0[a]	18.4[a]	80.1	329.2
1952	169[a]	132.2[a]	78.2	0.230	71.5[a]	60.2[a]	84.1	455.0
1957	412[a]	274.2[a]	66.6	0.424	172.6[a]	134.6[a]	77.8	490.9
1962	582[b]	350.0[b]	60.1	0.497	–	–	–	–
1965	737[b]	440.0[b]	59.7	0.589	–	213.0[b]	–	484.1
1970	1050[b]	615.0[b]	58.6	0.735	–	298.0[b]	–	484.6
1974	1459[b]	865.0[b]	59.3	0.951	–	420.0[b]	–	485.5
1975	1598[b]	945.0[b]	59.1	–	–	458.0[b]	–	484.7

[a] Wei-ta ti shih-nien [Great ten years], pp. 129-131.

[b] PRC: Handbook of Economic Indicators, C.I.A. Research Aid, ER 76-10540, August 1976, tables 17 and 19.

[c] Population statistics for metric ton per capita calculations are drawn as follow: 1949, 1952 and 1956 figures from the official, T'ung-chi kung-tso [Statistical Work]. no. 11, 1957; figures for other years are mid-year population, except that of 1974 which is estimated January population, from John S. Aird. Population Estimates for the Provinces of the PRC, 1953-1974 U.S. Bureau of the Census, 1974, pp. 22-23, tables I and II. For want of consistency, no effort is made to employ a different set of population statistics for the year of 1975.

water transport, the railway is clearly the preferred mode for long and medium distance travel since it accounts for more than 70 per cent of all passenger turnover up to 1960 (table 25). It should also be noted that the average Chinese travelled more than three times as frequently in 1960 compared to 1949. However, surprisingly, with the expanding network, the average passenger travelled over an ever shorter distance during the period under discussion, from a maximum of 144 to only 118 passenger-kilometers.

Areally, passenger traffic as reflected by the national passenger train timetable in 1956 was concentrated generally along a north-south corridor comprising the Harbin-Shenyang, the Shenyang-Peking, and the Tientsin-Shanghai lines, particularly over the central segment between Shenyang and Peking. Figure 16 also pinpoints some of the bottlenecks in the mid-1950's, for example, that to the north of Wuhan on the Peking-Wuhan line, and the section between T'ungkuan and Hsian on the Lung-Hai line. Except on the major trunk routes, passenger traffic on all other lines was at best insignificant. This pattern was almost completely changed by 1963. Although the Peking-Shenyang line remained the axis of the system, the Peking-Wuhan and Wuhan-Canton lines became the most important north-south passenger route, and the oldest east-west route, the Lung-Hai line which was then connected with the Lanchou-Hsinchiang line, became the prime mover of east-west routed passengers between Hsuchou and at least as far as Lanchou if not further (fig. 17). In the south, lines constructed during different periods now formed a second east-west route from Shanghai through Chuchou to Liuchou and Nanning and carried an increasing number of passengers. In addition, traffic in the industrial Northeast, in North China between Peking and Paot'ou and between Chinan and Ch'ingtao was clearly very substantial. Traffic in Ssuch'uan province, where the subnet was connected finally to the main system, also grew rapidly.

Railways and the Contraction of Space

There is no doubt a profound relationship between the Chinese space-polity and the development of modern transportation and communications systems.[1] In the context of China, where terrain is difficult and navigable waterways limited, and where economic conditions either delayed the development or restricted the functions

---

[1]Joseph B. R. Whitney, China: Area, Administration, and Nation Building, pp. 45-49; and Alan P. L. Liu, Communications and National Integration in Communist China, pp. 12-14.

TABLE 25

RAILWAY PASSENGER TRANSPORT PERFORMANCE, 1949-1960

	Total M	Passenger Carried Railway only			Total Bil/km	Passenger Turnover Railway only		
		M	%	ave. trips per cap.[b]		Bil/km	%	ave. Pass/km[a]
1949	135	103	76.3	0.240	15.4	13.0	84.4	126.25
1950	201	157	78.1	0.284	23.9	21.2	88.9	135.36
1951	220	160	72.9	0.284	26.9	23.1	85.6	143.74
1952	240	164	68.0	0.285	24.7	20.1	81.3	122.68
1953	350	229	65.3	0.389	34.8	28.2	80.9	123.22
1954	367	233	63.4	0.387	36.9	29.5	79.9	126.56
1955	361	208	57.6	0.338	35.2	26.7	76.0	128.55
1956	496	252	50.8	0.401	46.4	34.4	74.1	136.47
1957	623	313	50.2	0.484	49.5	36.1	73.0	115.57
1958	736	346	47.0	0.527	57.1	40.9	71.7	118.37
1959	–	481[a]	–	0.716	78.8[a]	56.7[a]	72.0	118.00
1960	–	530[a]	–	0.774	86.9[a]	62.5[a]	71.9	118.00

SOURCE: All figures and calculations except otherwise noted are from Wei-ta ti shih-nien [The great ten years] p. 133.

[a]Figures directly taken from Wu Yuan-li, The Spatial Economy of Communist China (New York: Frederick A. Praeger, 1967), p. 176, table 8-1.

[b]Population statistics for average trips per capita calculations are drawn as follows: 1949-1956 from T'ung-chi kung-tso, 1957; other years from John Aird, Population Estimates.

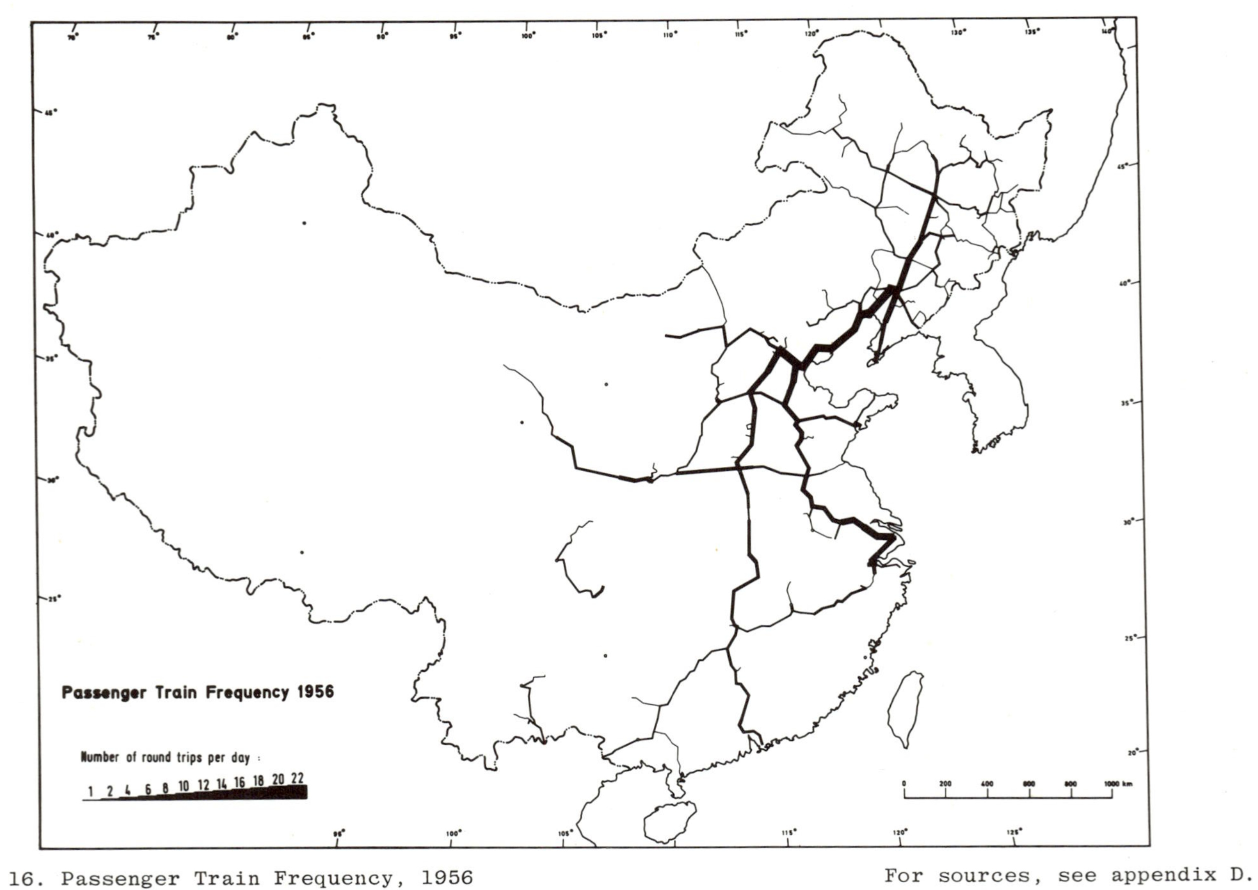

Fig. 16. Passenger Train Frequency, 1956 For sources, see appendix D.

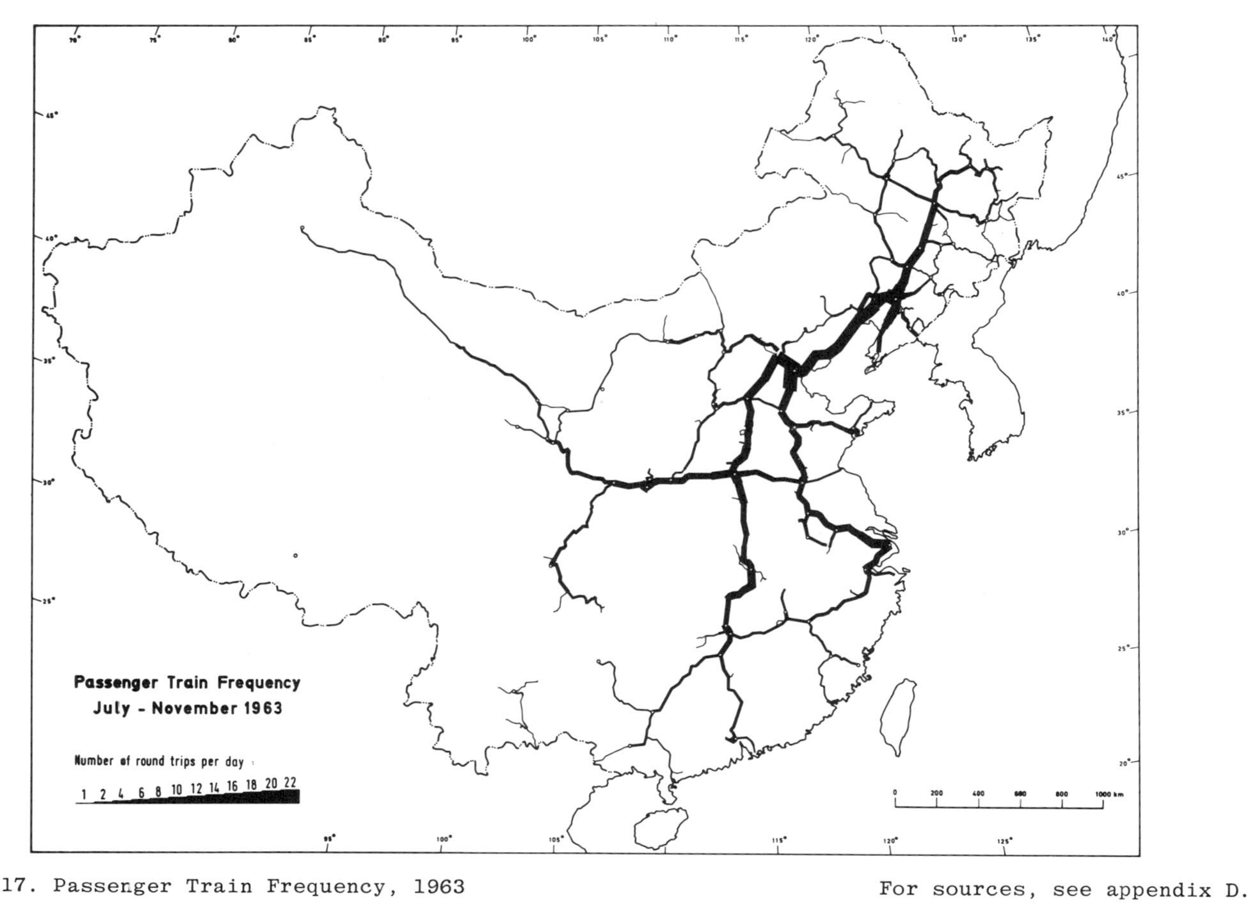

Fig. 17. Passenger Train Frequency, 1963

For sources, see appendix D.

of other modern means of transportation such as road and civil aviation, the railway has been vital in modifying traditional spatial relations. Construction of new railways and improvements on existing lines both have the effect of contracting the functional space of China. An analysis of the shortest physical distances by rail between selected pairs of cities for the years 1956 and 1963 clearly demonstrates this. Compared to 1956, a general contraction of space was experienced by all railway cities in China in 1963, as a result particularly of the improvement of the existing lines, with the only exception of Hsian, and to some extent Lanchou. What is more striking is the tremendous spatial reduction brought about by new construction, for example, the Paot'ou-Lanchou line, to the interior and frontier cities, namely, Lanchou, Paot'ou, Tat'ung, and Manchouli (fig. 18).

More important has been the contraction of time-distance resulting from the introduction and expansion of modern transportation systems. Indeed, between 1907 and the 1930s, time-distance generally contracted by 97 per cent within the territory served by railways.[1] Even after 1949, the vast but varied space of China continued to shrink with improvements in the system of circulation. Figures 19 and 20 show not only that an increasing proportion of the territory has been brought into contact by railways between 1956 and 1963, but also that most of the important cities in the Northeast, North, Central, and part of East China could be reached from Peking within twenty-four hours by 1963. Specifically, time-distance between Peking and Canton, for example, was shortened by nearly 35 hours or about 43 per cent compared to pre-1949 train journeys,[2] while the 4,099 kilometer distance between Shanghai and Urumchi and the 4,499 kilometer distance between Canton and Manchouli can be conquered, by express trains, in only 87 and 90 hours respectively.[3]

---

[1]In areas served by motor vehicle transportation, time-distance contracted by 83 per cent. See Joseph B. R. Whitney, China: Area, Administration, and Nation Building, p. 47; and Yang Ch'ing-k'un, "Chung-kuo hsien-tai k'ung-chien chu li chih shu tuan" [The construction of Space in modern China], Ling-nan hsueh pao. 10 (December 1949): 154.

[2]Yang-cheng wan-pao [Canton Evening Post], 10 October 1964, Canton.

[3]China Reconstructs 14 (July 1965); and Hsin-min wan-pao [New People's Evening Post] 11 April 1965, Shanghai.

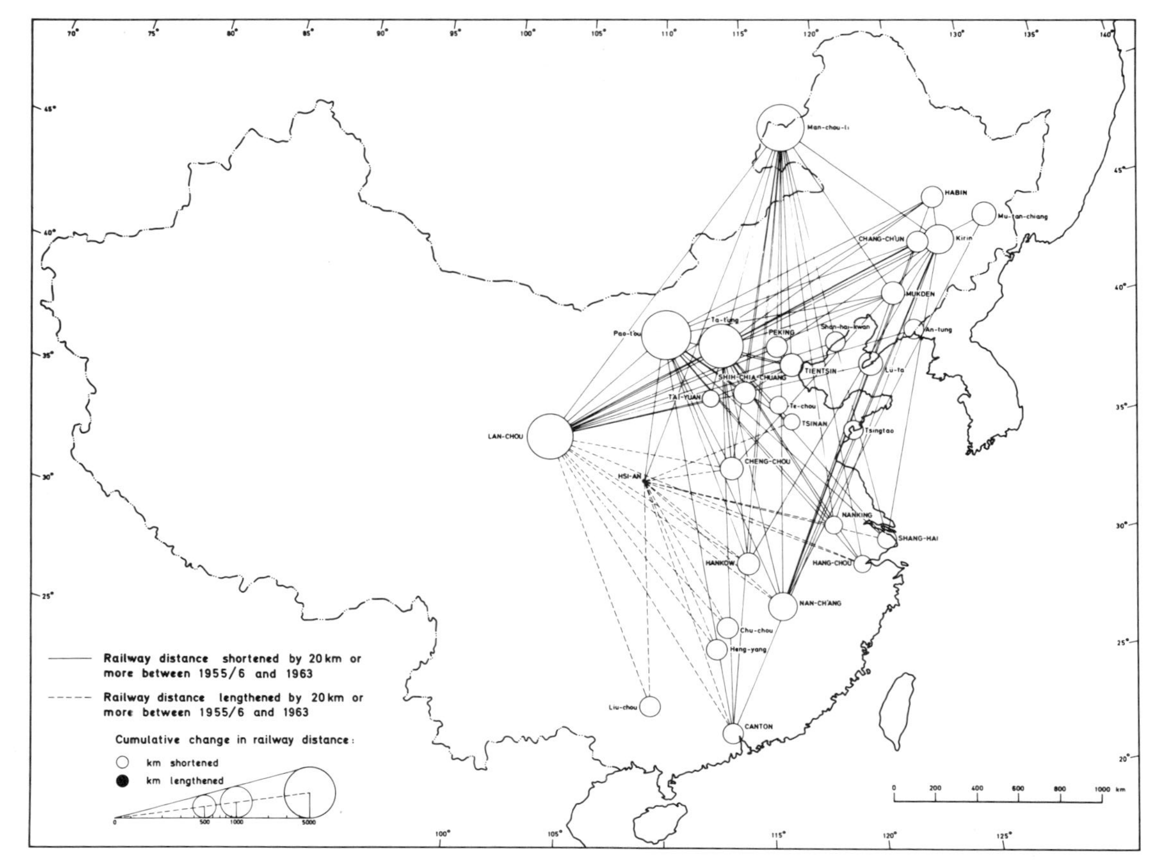

Fig. 18. Railway Distance of Selected Pairs of Cities, 1956-1963 For sources, see appendix D.

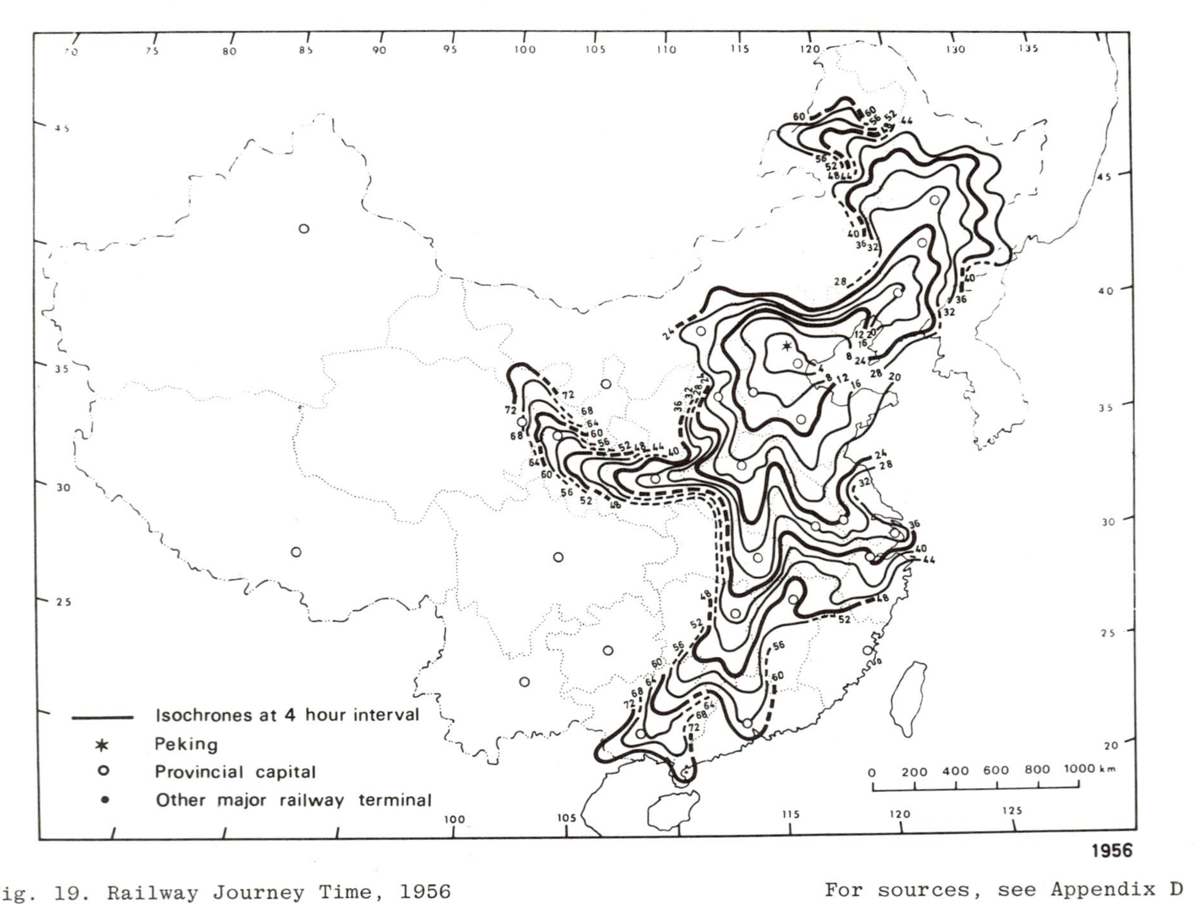

Fig. 19. Railway Journey Time, 1956

For sources, see Appendix D

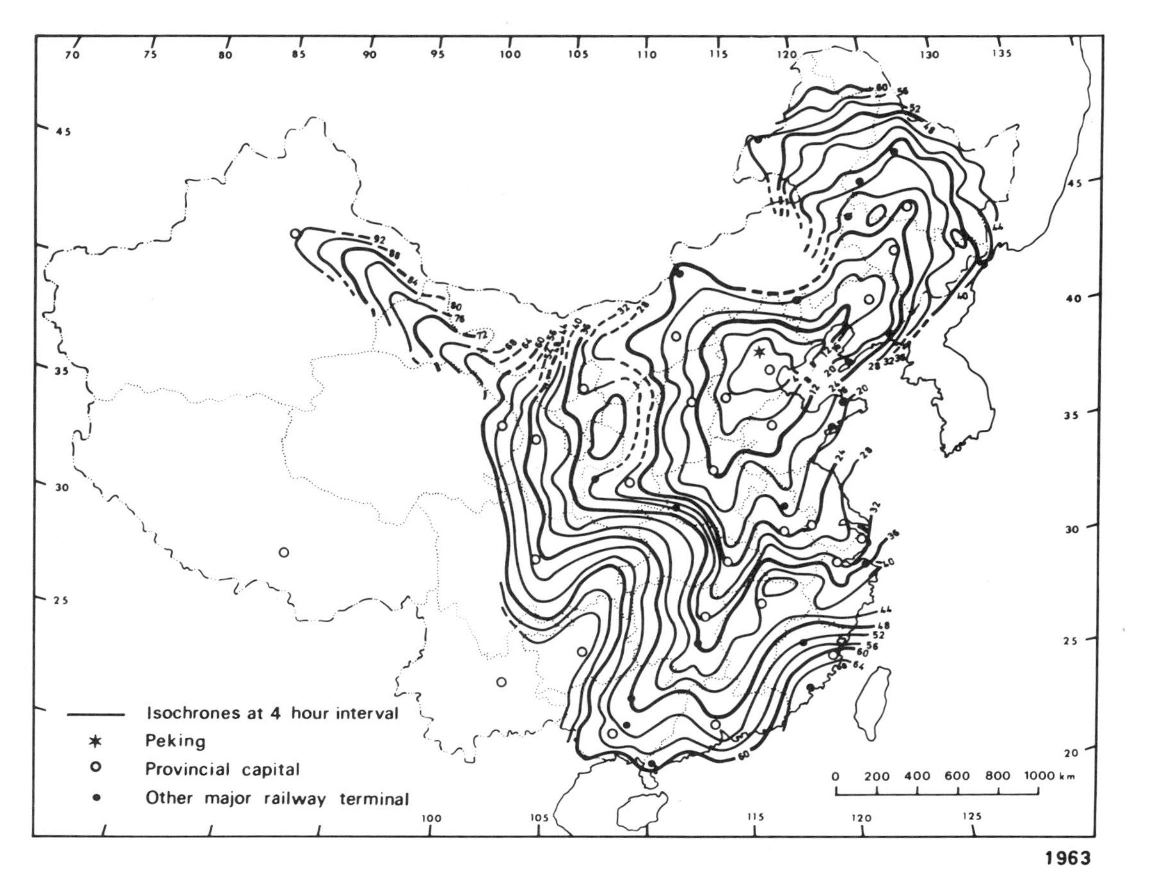

Fig. 20. Railway Journey Time, 1963

For sources, see appendix D.

Summary

Of all modern systems of transportation, the railway has a uniquely important function to play in Communist China at least up to the early 1970s. That this is so can be evidenced by several facts. First, the total route length of the network has been expanding at an annual rate of over 1000 kilometers of repaired and new lines, in spite of economic and political crises. This compares favorably with the speed of construction during any previous period in China's railway history. Second, the railway not only receives very substantial state allocations of investment funds, ensuring its steady development and expansion, but also is probably the most profitable component of the transport, postal, and communications enterprises in China. Third, modern transportation including the railway system is so important to China that even at times of economic and financial stringency, its supplies of equipment in terms of imports appear to have always been guaranteed. Fourth, the railway is undoubtedly the basic system of transportation in Communist China for it carries the bulk of freight and passenger movements, and this position is not likely to change for some time to come.

CHAPTER V

RAILWAY PATTERNS AND NATIONAL GOALS: AN EXPLORATION

Communist Ideology and Transportation Development

Mao Tse-tung Thought and Development Contradictions

In a consideration of the relationship between political ideologies and the space-polity of Communist China, nothing perhaps is more relevant than a review of the development of Marxism-Leninism, or dialectical materialism, in China through an examination of the so-called Mao Tse-tung thought.[1] Not only is China building communism in her own way, but also Mao Tse-tung has been instrumental in the Chinese Communist Revolution as well as in its post-Revolution socialist construction. The point has often been made that the Chinese revolution diverges from the Soviet model by its emphasis on the importance of the peasants and the countryside, for even Mao himself has pointed out that, "In Russia, the revolution developed from the cities to the countryside, while in our country it developed from the countryside to the cities."[2] Yet the Chinese divergence must be explained in more subtle terms, since even after the establishment of the communist government Mao continued to actively seek alternative development strategies superior to those employed by Russia and other East European countries.[3]

There is no doubt that in the interpretation of dialectical materialism, Mao has made important contributions. Specifically, in dialectics, Engels, following Hegel, emphasizes that all events and

---

[1] Steve S. K. Chin, Mao Tse-tung shih-hsiang: sing-shih yu nei yung [The thought of Mao Tse-tung: form and content], Centre of Asian Studies (Hong Kong: University of Hong Kong, 1976).

[2] Mao Tse-tung, "Talk to Music Workers," 24 August 1956, as quoted in Mao Tse-tung Unrehearsed, ed. Stuart Schram (London: Penguin Books, 1974), pp. 84-90.

[3] Mao Tse-tung, Mao Tse-tung shih-hsiang wan-shui [Long live the thought of Mao Tse-tung] (n.p., 1969), p. 222.

matters develop dynamically according to three parallel fundamental laws, namely, the mutual transformation of quantity and quality, the interpenetration of opposites, and the negation of negation. Although Hegel's idealistic dialectics have been criticised by Marx and Engels and transformed into materialistic dialectics, the triad (thesis - antithesis - synthesis) formula remains popular. However, the existence of a final and presumably static stage of "synthesis" obviously defeats the assumption in dialectics that the motive force in the continuous dynamic development of events and matters is found in the struggle between contradictory opposites. Mao therefore rejects the principle of the negation of negation on the grounds that every historical phenomenon was simultaneously or successively affirmation and negation,[1] and singles out the law of the (temporary and conditional) unity of contradictory opposites as the only fundamental law, arguing that the other laws are but subordinate ones. Insofar as this fundamental law is accepted, events and matters which develop as a result of the struggle between contradictory opposites can be better understood if the method (or viewpoint) of "one divides into two" is employed.[2] Mao therefore further argues that the dynamic development of events and matters depends primarily on internal rather than external causes.[3] For Mao, contradictions are not merely, as for Hegel and Marx, the motor of change; they are the inherent nature of all events and matters.[4] In the resolution

---

[1]Stuart Schram, ed., Mao Tse-tung Unrehearsed, Introduction, p. 26.

[2]This on the one hand effectively refutes the popular formula in materialistic dialectics, and on the other improves its consistency with Marxism-Leninism in that the state of unity of contradictory opposites not only is temporary and conditional but also begets new contradictions and therefore struggles. This explains why Stalin and his successors regard the communist society as an ideal society where there are no internal contradictions, while Mao believes that even after socialist revolution the road to socialism and communism will be a long and difficult one, being full of internal and external contradictions the solution of which is through continuous revolution. Moreover, in conformity with the universality of dialectics, even socialism and communism will develop dynamically, will be only some stages in the endless evolution of social history, and will be succeeded by yet other forms of society.

[3]For detailed analysis and discussion of Mao Tse-tung's development of dialectical materialism, see Steve S. K. Chin, Mao Tse-tung; and C. K. Leung, "A Review of the Thought of Mao Tse-tung: Form and Content by Steve S. K. Chin," The China Quarterly, no. 68, December 1976, pp. 845-848.

[4]Stuart Schram, ed., Mao Tse-tung Unrehearsed, Introduction, p. 25; and Steve S. K. Chin, Mao Tse-tung.

of contradictions, Mao uniquely emphasizes transformation rather than annihilation, and thus the use of commonality in the process of struggle for the promotion of interpenetration or mutual transformation. Mao clearly believes that contradictions must be resolved in the transformation of opposites.

Communism envisages a classless society without exploitation, and the transition from capitalism to communism must accomplish the elimination of the differentials between the bourgeoisie and the proletariat, between the mental and manual workers, between the (industrial) workers and peasants or more generally between industry and agriculture. In spatial terms, such differentials may be summed up by contradictions between the city and the countryside, or between the core and the periphery. In searching for an alternative developmental strategy for socialist construction, especially one that would appeal to the desire for "more thoroughly, faster, better, and more economically building socialism," Mao early in 1956 highlighted ten major contradictions in the Chinese context.[1] Of these ten contradictions, six[2] are essentially geographical contradictions in that they are rooted in the consideration of the regional differences, whether economic, physical, political, or social, that existed in China. To resolve these contradictions, Mao points out that of the two alternatives in developing heavy industry, the one that emphasizes agriculture and light industry and thus the satisfaction of the needs of the population is clearly the superior one and one that China should follow; that whereas the policy of developing industry in the interior is right, and should continue, industry in

---

[1]Mao Tse-tung, "On Ten Major Relationships," in Mao Tse-tung, Mao Tse-tung shih-hsiang wan-shui, 1969, pp. 40-59. The ten major relationships were apparently formulated earlier in 1955, presumably formally presented at a meeting of the Politburo in April 1956. See also Mao Tse-tung, ibid. p. 163. These relationships have been aptly referred to as dialectics of development by Stuart Schram, ed., Mao Tse-tung Unrehearsed, 1974, p. 24.

[2]Namely, these are contradictions (1) between industry and agriculture and between heavy industry and light industry; (2) between industries in the coastal regions and those in the interior; (3) between economic construction and defence construction; (4) between the State, the production unit, and the individual producer; (5) between the Center and the regions; and (6) between the Han nationality and the minority nationalities. See Mao Tse-tung, Mao Tse-tung shih-hsiang wan-shui, pp. 40-59.

the coastal regions must also be appropriately developed in order to better support and sustain the development of industry in the interior and the restoration of regional equilibrium; that the best policy to improve national defence is not to increase military or administrative expenditure but to increase economic construction; that the best way to safeguard the interests of the State is to promote the interests of the individual producer and to allow greater accumulation by the production unit; that to strengthen the leadership of the Center, the interest and independence of the regions must be taken care of; and that the contradiction between the Han and the minority nationalities must be resolved primarily through anti-Han chauvinism.

Mao's views on these major contradictions and his approaches to their resolution as outlined above were eventually officially adopted as guidelines for national development in January 1958,[1] though they must have long found their way into the formulation of the development plans such as the five-year plans.

Spatial and Regional Contradictions

If Mao is right, his ideologies must have developed from and be suited to the environs of the Chinese society, for Mao himself has asserted that people's correct ideologies can only come from practice, particularly from production struggle, class struggle, and scientific experimentation in human society. In short, the social existence of man determines his thoughts and ideologies.[2] Developmental contradictions must be resolved within the given (political) territory and (physical) space, since modern nation-states not only have to operate within the confines of their territories, but also have to adjust to the given physical space which is often differently endowed. Mao's conceptualization of Chinese national development can be regarded as a process of spatial integration. Yet integrative processes reflect not merely the quality of space as an objective reality but also political ideologies which are frequently shaped by reality.

---

[1]Ibid., p. 278.

[2]Mao Tse-tung, "From Where Man's Correct Ideologies Come?," in Mao Tse-tung, Mao Tse-tung chu-cho hsuan-tu [Selected readings of Mao Tse-tung's writings] (Peking: People's Press, 1969), pp. 383-384.

The Political Space: As a pre-modern state, China under the Ch'ing dynasty had a national ecumene, extra-ecumenical areas, and a number of dominions and protectorates on the periphery. The Manchu Empire of the Ch'ing dynasty therefore stretched from far beyond the estuary of the Amur River in the north to Borneo and the Malacca Straits in the south, and from the Ryukyu islands in the east to eastern Turkistan beyond Lake Balkhash.[1] These territories, however, began to break up surprisingly soon after the establishment of the Empire, for conflicts between Russia and China after 1689 soon led to the Bura-Kyakhta Treaties in 1727 which ceded to Russia the vast area northwest of the Amur River.[2] Thereafter, China's national territory continued to shrink during the Ch'ing dynasty, first through either annexation by other powers or the independence of her dominions and protectorates, and later by an encroachment upon her extra-ecumenical and even ecumenical areas. China's continental, or the inner Asian,[3] frontiers were among the earliest and most persistent areas of troubles. Thus, China traditionally looked towards her continental frontiers with apprehension until the beginning of the Nineteenth Century when the sea powers began to encroach upon her coastal areas. From the 1690's onwards, China's national territory contracted from at least 13 million square kilometers[4] to roughly 11.165 million square kilometers in 1911 under the Republic of China and to only 9.6 million square kilometers in 1949 under the People's Republic of China.

The Physical Space: Today, China's territory of some 9.6 million square kilometers measures 5,200 kilometers east-west and 5,500 kilometers north-south. The vertical diversity too is impressive: the peak of the Jolmo Lungma on the Himalayas rises to

---

[1] Fang Ch'iu-wei, Chung-kuo pien-chiang wen-ti shih-chiang [Ten lectures on China's frontier problems], (Shanghai; Motor Press, 1937), pp. 4-5 and Table; J. B. R. Whitney, China: Area, Administration, and Nation Building; and Albert Herrmann, An Historical Atlas of China (Chicago: Aldine, 1966), revised edition, ed. Norton Ginsburg.

[2] Fang Ch'iu-wei, Chung-kuo pien-chiang wen-ti shih-chiang; and Violet Conolly, Siberia Today and Tomorrow (London: Collins Press, 1975), p. 30.

[3] Owen Lattimore, Inner Asian Frontiers of China (New York: American Geographical Society, 1940).

[4] Excluding areas of all dominions and protectorates, and including only major losses to Russia, etc., between 1680's and 1911. See Fang Ch'iu-wei, Chung-kuo pien-chiang wen-ti shih-chiang, table 2, between pp. 4 and 5.

8,882 meters, whereas part of the Turfan Depression in Hsinchiang drops to 154 meters below sea level. Generally, altitude decreases from west to east, and China's physical space may be likened to a three-step system. The first and highest step is the well-known "roof of the world', i.e., the Ch'inghai-Tibet plateau. With an average height of 4,500 meters and a total area of 2,300,000 square kilometers, it is the world's largest and highest plateau area. North and east of the Ch'inghai-Tibet plateau and west of a line joining the eastern slopes of the Greater Khingan and the eastern limits of the Yunnan-Kueichou plateau in the southwest is the second step. This area of central-west and northwestern China has an average altitude of about 1,000-2,000 meters or lower. On the second step are found not only most of China's deserts but also pasture lands into which Chinese peasants traditionally tried to penetrate. The third step roughly corresponds to the southeastern third of China, with a general elevation of below 500 meters, and containing most of the fertile plains in China, e.g. the North China plain, the middle and lower Yangtzu plains, and the Northeast plain (fig. 21).

Regional Socio-Economic Differences: Generally, therefore, hostile land in terms of extremely high altitude, desert and marshy environments, or earthquake-prone areas, etc., decreases from the first through the second to the third steps. Conversely, fertile agricultural land and rainfall decrease westwards and northwestwards. This pattern is reflected in the distribution of population and economic activities. Thus population has been concentrated on the southeastern half of China. Even in 1953, the year when the first nation-wide census was undertaken by the newly established Communist government, such imbalance was hardly changed. Figure 22 shows that as much as 80 per cent of the population could actually be found in about one-third of the total area, and half of the land carried no more than 6 per cent of the population. Similarly, operating railways in terms of route length, though not necessarily an entirely satisfactory indicator of economic activities, were even more concentrated in that less than 2 per cent was actually found in nearly half of the area.

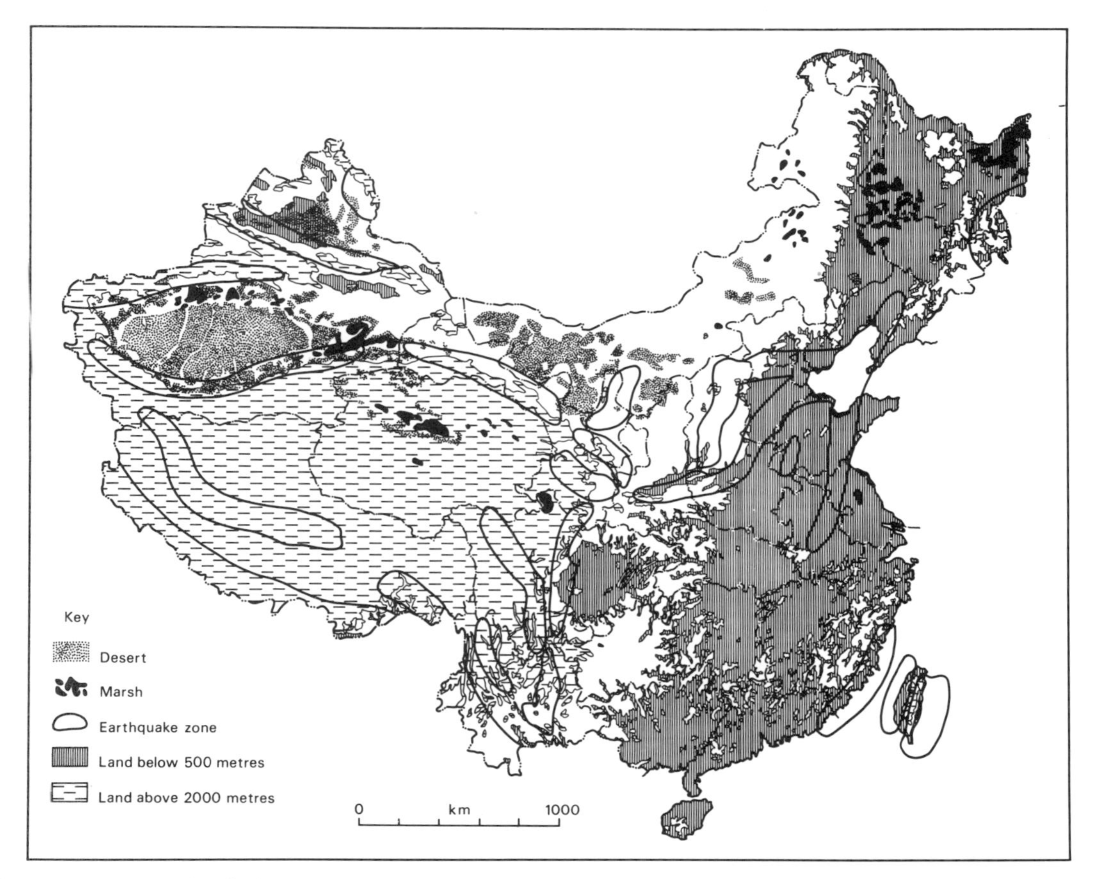

Fig. 21. The Hostile Land of China

For sources, see appendix D.

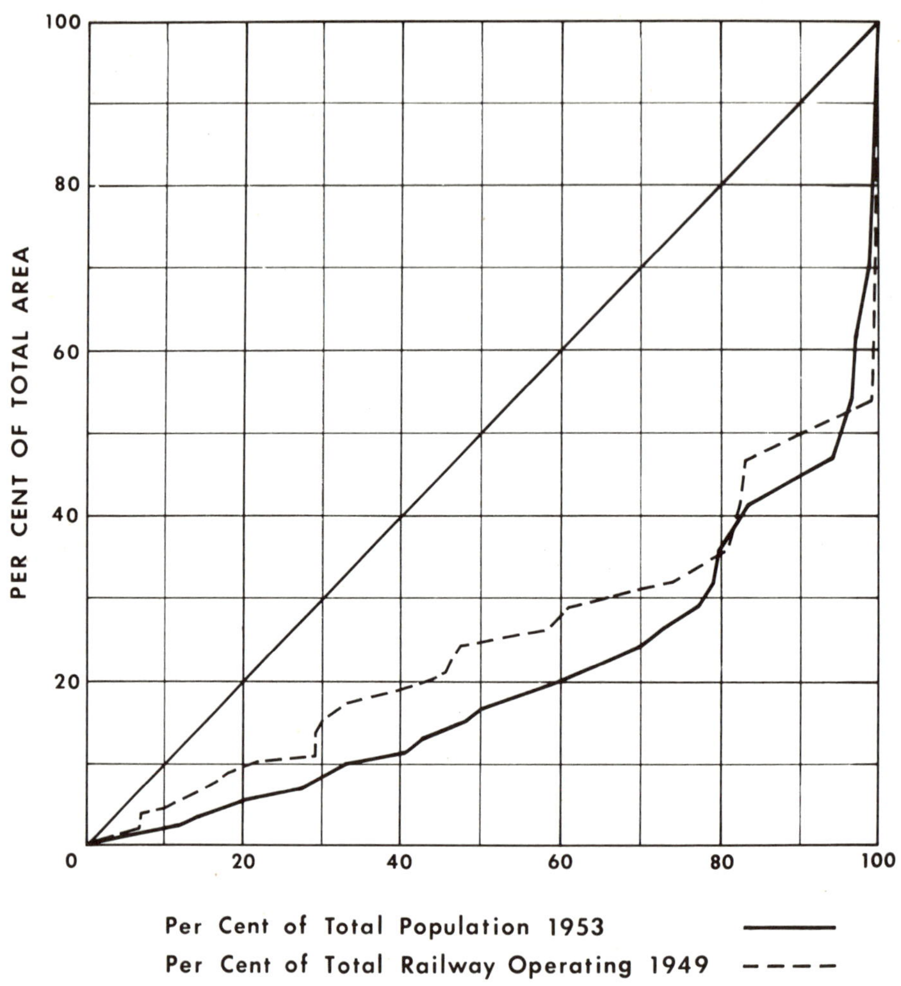

Fig. 22. Distribution of Population and Operating Railway

Transportation Planning

That the development of communications and transportation may be conditioned by the physical landscape is evidenced by the fact that since 1949 extensive researches have been carried out on the functions of transportation and the relationship between physical conditions and transportation development.[1] Indeed, exhaustive studies have been made on the demarcation of physical regions in terms of their natural conditions for the development of land transportation. Such studies provide important data for the selection of planning parameters, such as maximum permissible gradients for, or volumes and costs of, railway construction work, to be used in different parts and over various kinds of terrain in China.[2]

The importance of communications and transportation in China's national economic construction has certainly also been fully recognised. Not only that the procurement, allocation, and sales of consumer goods in China are planned, but that the transportation of such goods must be realized on a planned basis. Moreover, national economic construction requires voluminous materials and equipment which must be transported. According to available statistics, for every 100 million yuan of investment in basic construction, there will be as much as 500,000-600,000 tons of building materials such as cement, bricks, rocks, sand, etc., passing through the railway system alone.[3] Transportation therefore has long been regarded as

---

[1]For example, Yang Wu-yang, "An Assessment and Regional Demarcation of the Natural Conditions for Land Transportation in China", Ti-li hsueh-pao [Acta Geographica Sinica], 30 (4) (December 1964): 301-318; Joint Publication Research Service, "An Analysis of the Relationship Between Transportation and the Location of Extractive Industries." JPRS no. 29013. From Ti-Li [Geography], no. 5, September 29, 1964, pp. 207-210; Chang Kuo-wu and Chang Chih, "An Examination of the Development, Location, Support to Agriculture of Local Railways in our Country," Ti-Li, no. 1, 1964, pp. 5-10; Joint Publication Research Service, "A Study on the Topography of a Desert Region for the Selection of a Railway Route." JPRS no. 33258. From Ti-Li [Geography], no. 4, July 27, 1965, pp. 167-170; Chao Sheng-chuan and Pan Pei-wen, "The Threat of Windblown Sand to Railways and Its Prevention," Ti-Li, no. 1, 1965, pp. 13-17; Chang Wu-tung and Wu Cho-liang, "On the Location of Transportation Network in Regional Planning", Ti-Li, September 1961, pp. 216-220.

[2]Yang Wu-yang, "An Assessment and Regional Demarcation."

[3]Lei T'ing and Liang K'uang-pai, "The Role of Transportation in Our National Economy," Ching-chi yen-chiu [Economic study], no. 2, February 1965, pp. 39-43.

the "vanguard" in the development of the national economy,[1] and must therefore advance before all others. While insisting on the principle of "walking on two legs", the Party clearly lays down directives and policies for the planning and development of transportation and communications. Generally, the planning and development of transportation and communications must proceed in advance of, though must be suited to, needs, must be sighted on the long-term future though started on the more immediate (needs), must be self-reliant and thus make use of local resources, must employ indigenous as well as foreign methods so as to improve steadily, and must give priority to connection over flow (volume) and thus line over network.[2]

Railway Patterns and the Chinese Space-Polity

Network Structure and Emphasis

The structure of various network may be analysed in terms of their topological or geometric components.[3] In topological analysis, networks are first transformed into graphs, each being an array of points connected or disconnected by lines. Since there is no concern with the actual length or orientation of the lines nor whether they are straight or curved, topologocal analysis may reveal common structures of apparently unlike networks. Topologically, places or points connected or to be connected are referred to as vertices (v) or nodes, connecting routes or lines as edges (e) or links, while a path is a collection of routes linking a series of different places. The length of a path is, in topological terms, the number of routes in it. The topological distance between two places therefore is the length of the shortest path joining them. A graph (G) is the entire

---

[1]According to, "Speed Up the Modernization of Transportation and Communications So As To Realize the Will of Chairman Mao," People's Daily, 1977, November 6, Mao had a grave concern for transportation and communications in China. Early in the revolutionary war period, Mao emphasized that the development of transportation and communications was an important precondition for the victory of the revolutionary war; since liberation, he subtly discussed the relevance of transportation and communications to the consolidation of proletarian dictatorship and its role in the national economy and defence construction.

[2]Chang Wu-tung, et al., "On the Location of Transportation Networks."

[3]William L. Garrison and D. F. Marble, The Structure of Transportation Networks; Karel J. Kansky, Structure of Transportation Networks; Peter Haggett and Robert Chorley, Network Analysis in Geography; and Ronald Abler et al., Spatial Organization.

network under study, whether or not it consists of non-connecting subgraphs (p).

Topological indices are often derived as indicators of national development in terms of economic achievement in general and technological achievement in particular. In a non-cross-national study such as this, time-series topological indices, in fact, can be very useful in depicting the evolution of a national network as well as the directions of the evolutionary process.[1] Table 26 presents the results of an analysis of the evolving Communist Chinese railway network of selected years in terms of topological indices. It will be seen that the extremely fragmented network in the mainland of 1949 (p=11 i.e. there were eleven separate unconnected systems) was rapidly linked up through the Rehabilitation and the First Five-Year Plan periods, though its complete integration was not achieved until

TABLE 26

TOPOLOGICAL INDICES OF RAILWAY NETWORKS IN MAINLAND CHINA, 1949-1975*

	e	v	p	δ	μ	α	β	γ	η	ι	π
1949	91	91	11	16	11	0.062	1.000	0.341	188.32	54.58	1.288
1952	129	121	8	19	16	0.068	1.066	0.361	174.84	49.79	1.621
1957	197	182	3	30	18	0.050	1.076	0.364	145.58	40.91	2.736
1963	254	234	2	33	22	0.047	1.085	0.365	134.16	37.49	3.546
1970	313	278	1	32	36	0.065	1.126	0.378	130.11	36.04	7.740
1975	354	311	1	35	44	0.071	1.138	0.382	124.24	34.52	8.057

*Excluding subnets of Hainan Island and Taiwan.

Note: Measures of network structure are constructed predominantly as ratios between the whole system and its individual elements. Only two measures, abstracted directly from graph theory (the cyclomatic number, μ, and diameter, d, of networks) are not ratio measures.

The Cyclomatic number (μ) (or first Betti number) is an arithmetic comparison between individual elements of the system (μ=e-v+p). In connected graph G, the cyclomatic number is equal to the maximum number of fundamental circuits. Thus any disconnected graph (p greater than 1) or any tree-like graph has a μ equal to zero. Highly connected graphs have higher μ values. It has been found that the cyclomatic number is a useful measure of the spatial structure of transportation networks, since higher μ values are generally found of transportation networks in highly developed countries.

[1]Edward J. Taaffe and Howard L. Gauthier, Jr., Geography of Transportation (Englewood Cliffs: Prentice-Hall, 1973), pp. 100-115

TABLE 26--Continued

The diameter (d) is the maximum number of edges in the shortest path between any pair of vertices ($d(G)=\max_{x} \max_{y} d(x,x)$). It is thus an index measuring topological length between vertices or the "extent" of a graph. Although the diameter is weak for measuring and comparing networks, in the present application as a time-series measure, the d does provide a useful means in gauging the development of a given network over time.

The Beta index (β) measures the relationship between two individual elements of a network, i.e., the number of edges over the number of vertices ($\beta=\frac{e}{v}$). Higher β values are produced by more sophisticated network structures, especially by those with a greater number of edges in relation to the number of vertices. While the lower portion of the β scale (from 0 to 1) differentiates between different types of branching or disconnected networks, values of 1 and above differentiates circuit networks with a maximum value of 3.00 for planar graphs.

The Alpha measure (α) is an adjusted form of the cyclomatic number, μ and it may be regarded as the ratio between the observed number of circuits and the maximum number of circuits in a given graph, $\alpha=\frac{\mu}{2v-5}$ For completely interconnected networks the α index will equal to 1, whereas networks with a decreasing degree of connectivity will have an index approaching zero. Zero α values will be obtained for networks with a cyclomatic index of zero. The α index is therefore independent of the number of vertices in the network. Seldom, though, would transportation network have an α value of 1, a completely interconnected graph, for it means serious redundancy.

The gamma index (γ) is a quotient of the observed number of edges to the maximum number of edges ($\gamma=\frac{e}{3(v-2)}$). The gamma index is bounded by 0 on the lower limit and 1 on the upper limit. The value of 1 describes completely connected networks, and lesser values indicate various degrees (or percentages) of connectivity. Similar to the α index, the γ index is independent of the number of vertices of a given network, and is very effective in time-series application.

The above are the most important topological measures of network structure. Other indices are generally weighted by either the total traffic flow (which may be the volume of freight, passenger, interaction, etc.), or the total kilometrage, or by the number of connections of any vertex of a given network under study. Three of these indices have been examined in the present study, namely, the eta, the iota, and the pi indices.

The eta index (η): both eta and pi are measures expressing relations between the transportation network as a whole and its routes as individual elements of the network. Thus the eta index is a ratio between the total kilometrage of the network and the observed number of edges, or a measure of the "average edge length" ($\eta=\frac{K}{e}$), here K is the total kilometrage. As a ratio, eta assigns ordered values to individual networks, so that networks having the same ratio between the whole network and all of their routes should have the same value. It is also apparent that as a network develops, its eta index is likely to decrease.

TABLE 26--Continued

<u>The iota index</u> (ι), a measure that considers three aspects of a transportation network: (1) structure, (2) length, and (3) functions, is the ratio between the transportation network as a whole and its weighted vertices ($\iota = \frac{K}{w}$). The network is represented by its total kilometrage, while the vertices are weighted either by their appropriate function or traffic flow. When it is weighted by function, the <u>iota</u> index can be interpreted as average distance per function. Decreasing <u>iota</u> values indicate increasing densities of functions in an evolving network.

<u>The pi index</u> (π), like π in geometry, measures relationship between the circumference of a circle and its diameter ($\pi = \frac{C}{d}$). Here C circumference is the total mileage of a given network, and d diameter is the total kilometrage of the network's topological diameter. Its lowest value of 1 increases with complexity of networks.

1970 (p=1). Whereas the number of edges (e) and vertices (v) both increased more than three-fold between 1949 and 1975, the actual ratio between the number of edges and the number of vertices (β) remained practically unchanged before 1970, or changed only marginally after that year. It will be recalled that, whereas β values between 0 and 1 identify branching networks, β values between 1 and a maximum of 3 indicate a range of densities of circuits. Thus, the β index points to the fact that the network since 1949 has been expanding rather than increasing in internal sophistication at least up to 1970 if not later. Similarly, in spite of the apparently great expansion in the diameter (d), or the "extent" of the network, and in the number of circuits, the cyclomatic number (μ), which together should normally characterise a growing system with improving connectivity and efficiency, both the quotient of the observed number of edges to the maximum number of edges (γ) and the ratio between the observed and the maximum number of circuits (α), powerful measures of the "absolute" development of an evolving network, either improved insignificantly or actually fluctuated within very narrow limits between 1949 and 1975. The resulting basic network configurations therefore at best must be classified as spinal networks, since the gamma index remained between 1/3 and 1/2, and the alpha values close to zero.[1]

Furthermore, as the pi value (π), a number expressing the relationship between the diameter and the circumference of a circle

---

[1]Ibid., pp. 105-111. Network configurations are classified into three groups: spinal, grid, and delta, based on the values of the γ and α indices.

represented by the total kilometrage of a network, apparently grew nearly seven times between 1949 and 1975, its most rapid growth was achieved between 1963 and 1970, a period when the expansion of the extent (d) of the network had actually stabilised. On the other hand, the average edge length (η) and the average distance per function (ι)[1] appeared to have improved only moderately though gradually between 1949 and 1975, the former from 188 kilometers in 1949 to 124 kilometers in 1975, and the latter from 54 kilometers to 34 kilometers over the same period.

These topological indices of the Chinese railway network clearly indicate an evolving system which has been expanding much faster than the increase in internal connectivity and sophistication. The emphasis of the evolutionary network appeared to have been the achievement of ever greater spatial extent and dispersion, i.e., territorial integration, rather than a maturing network compatible with an industrialising and growing economy, as would have been the case in other network development.

Nodal Accessibility and Areal Organization

When examining a transportation network for evidence of the areal organization of a territory, a geographer is not restricted to considering only the aggregate characteristics of the network, i.e., the topological values. He may examine the nodes themselves in terms of their functions and accessibility to the rest of the network in order to construct some pattern of areal dominance and competition among the nodes.[2] For this purpose, a network is represented by a matrix, which can be subjected to a number of measures of nodal accessibility. A practical procedure for measuring nodal

---

[1]According to Karel J. Kansky, Structure of Transportation Networks, pp. 25-28, the ι index, $\frac{K}{W}$, is designed for networks for which no traffic data are available, and is based on the intuitive assumption that the structure of a transportation network is a reflection of the network's traffic pattern. The ι index is calculated as follows: the numerator is the total number of kilometers or miles of a given network, the denominator is the sum of the network's vertices, weighted individually by two when vertices are the third and higher order intersections. (An nth order intersection is by definition an intersection with n rays).

[2]Alfonso Shimbel, "Structural Parameters of Communication Networks," Bulletin of Mathematical Biophysics 15 (1953), pp. 501-507, as quoted in Edward J. Taaffe and Howard L. Gauthier, Jr., Geography of Transportation (Englewood Cliffs: Prentice-Hall, 1973).

accessibility while eliminating redundancies from the computations has been devised by Shimbel, who computed a matrix D to indicate the (topological) distance of the shortest path between all pairs of nodes in a given network. Operationally, the matrix D is computed by successively powering the connection matrix C and noting after each iteration the occurrence of non-zero elements. The procedure is terminated when no zero elements can be entered into matrix D, i.e., when the power of the matrix is equal to the diameter of the network. The element entries in this final matrix D indicate the lengths of the shortest paths between all pairs of nodes and, summing across the rows, the vector elements measure the accessibility of a node to the entire network.[1]

Matrices representing the Chinese railway networks of the selected years have been similarly analysed, and results only of the years 1949, 1963, and 1975 will be presented here. It will be seen that the most accessible ten cities, those ranking 1 through 10 in this analysis, in 1949 formed an axis between Peking and Harbin, and that the Northeast as a region was by far the more accessible. Because of the damage to the north-south trunk line, the then Hankow-Canton railway, the South was pretty much cut off from the major national system (fig. 23). This pattern was gradually changed through 1952, when the Hankow-Canton line was restored, and 1957, when the railway network began to expand toward the Northwest and Southwest. By 1963, the accessibility surface in general was vastly extended toward the Southwest, the Northwest, and the Southeast, whereas the "core" formed by the ten most accessible cities reflected the shape of the expanding network and shifted southwestward to North China. Remembering that the Northeast had been, and still was, the most important industrial region and that the Southwest as well as the Northwest the least developed areas, this shift was significant (fig. 24). It would also be true to state that by 1963, most of the features of the present pattern of nodal accessibility had been established, since the 1970 and 1975 patterns appeared to have refined some of the earlier details only (fig. 25). Specifically, the accessibility surface in 1975 remained little changed, except in the extreme Southwest where the completion of the Ch'eng-Kun railway brought the area into the national system. Since 1949, the most accessible city shifted along a northeast-southwest corridor from Shenyang in 1949, through Chinchou in 1957, Tientsin in 1963, and finally to Shihchiaochuang in 1970 and 1975 (appendix C).

---

[1]Ibid., p. 133.

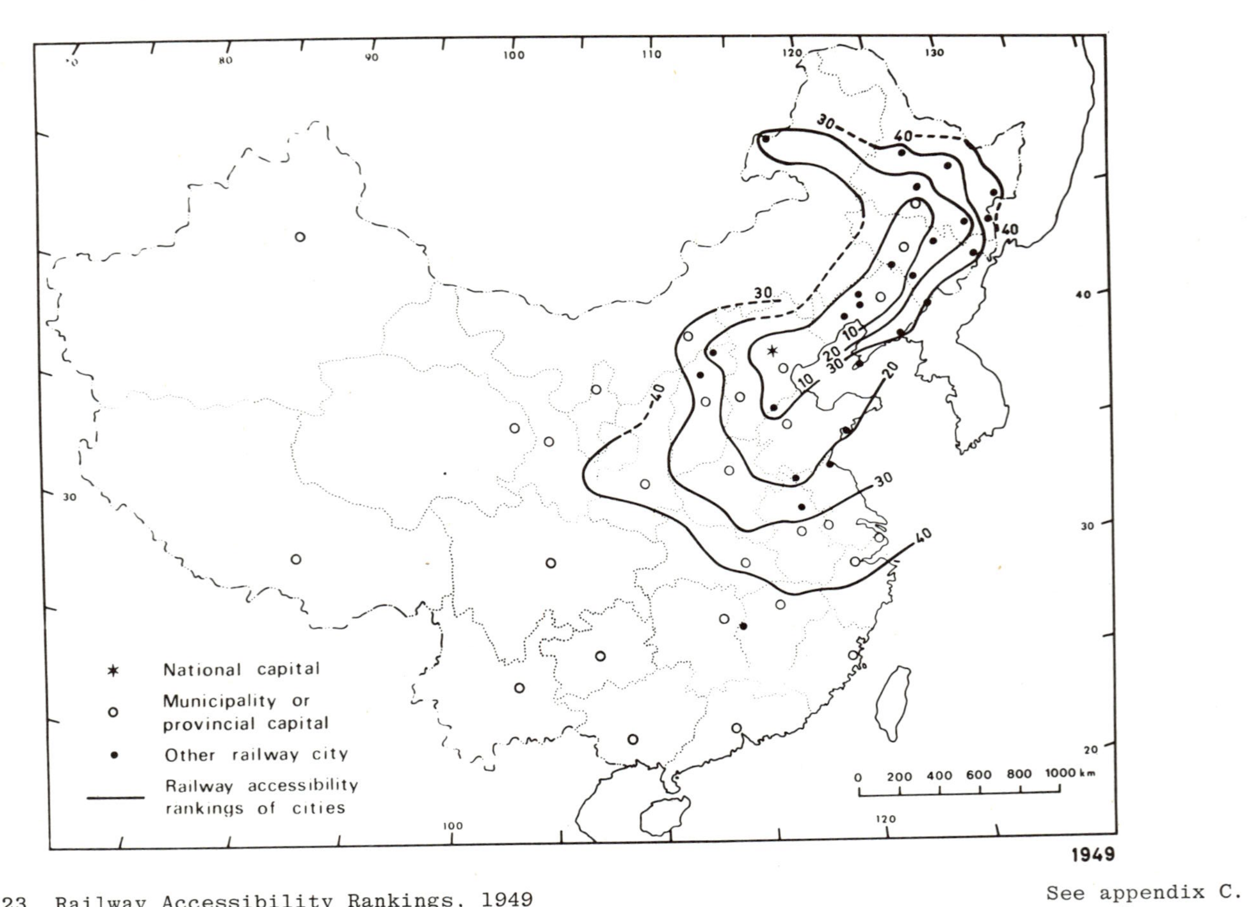

Fig. 23. Railway Accessibility Rankings, 1949

See appendix C.

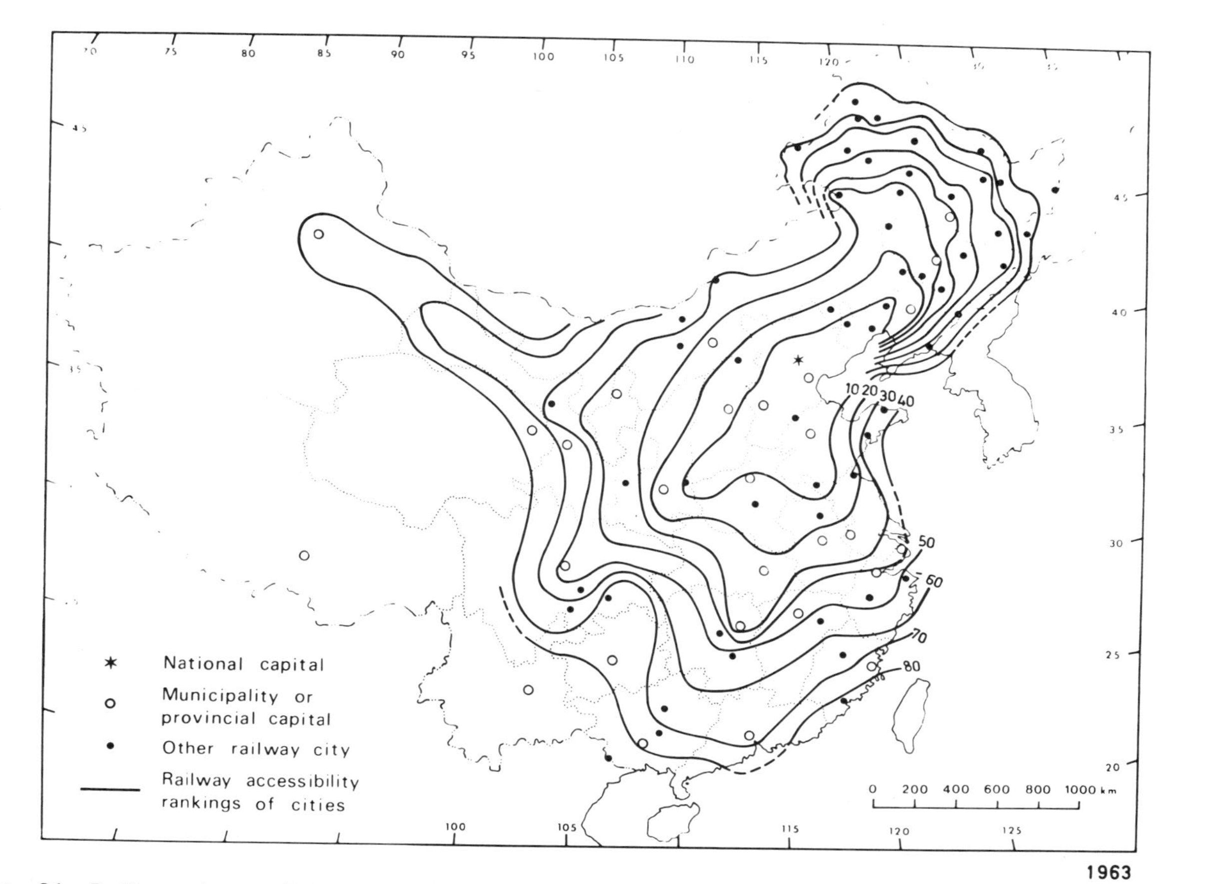

Fig. 24. Railway Accessibility Rankings, 1963

See appendix C.

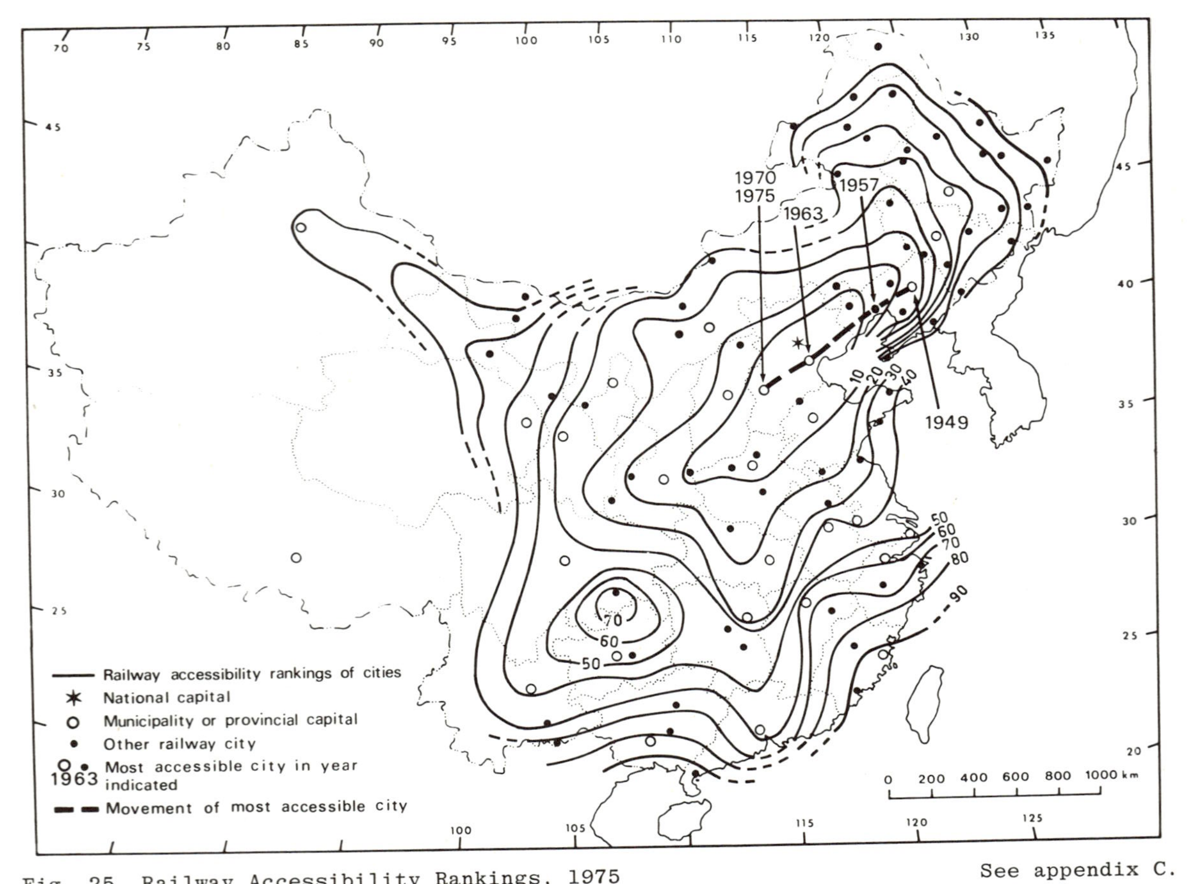

Fig. 25. Railway Accessibility Rankings, 1975

See appendix C.

Railway Connectivity and Areal Hierarchy

Purely economically motivated transport lines characteristically connect a point of production with one of consumption, or a hinterland with a port, or simply two productive areas. Politically motivated transport routes, however, may not show the same economic consciousness. In other words, the vertices joined by a transport route may not have dominant economic relationship. However, since transport routes are generally multipurposed, it will be extremely difficult to isolate political aspirations from economic ones. Yet politically motivated transport routes are certainly aimed at efficient areal organization, which to a large extent is determined by the connection or non-connection of the administrative centers and the presence or absence of some kind of areal hierarchy among these administrative centers. Specifically, in spite of the fact that the evolving network has a very low ratio between the observed and the maximum number of circuits (α), and shows an inclination toward dispersion rather than increasing internal sophistication, an efficient organization of political areas is more determined by the manner in which the administrative nodes or vertexes are connected by the route network, both at the lower level of the horizontal relationship among provincial capitals, and at the higher level of the vertical relationship between provincial capitals and the national capital, Peking.

A test is therefore designed on this basis. First, provincial and autonomous regional capitals are taken as the vertexes, and Peking as the administrative center. A province or autonomous region is considered to be connected with its neighboring province or region provided (1) that there is a railway linking up the two vertexes and (2) that this linking route does not pass through another vertex, though it may pass through the area of a third province or autonomous region. Any vertex is considered to be connected with the administrative center, Peking, as long as there is a continuous railroad between them, however circuitous the actual routing may be. At the lower level, i.e., vertex-to-vertex connection, the number of neighboring or adjacent vertexes not connected by a railway is expressed as a ratio to the total number of neighboring or adjacent vertexes of a given province or autonomous region. At the higher level, i.e., vertex-to-capital connection, non-connection is expressed by 0 and connection by 1. In both cases, the municipalities of Tientsin and Shanghai are not considered provincial level units for the present purpose for obvious reasons. The analysis is carried out for the years 1949, 1952, 1957, 1963, 1970 and 1975 in conformity with the previous analyses. The results are presented in table 27.

The analysis reveals several important features of the evolving railway network in China. In 1949, as expected, more than half (53.83%) of the provincial and autonomous regional capitals had no rail connection with Peking, however indirect, and the 26 provinces and autonomous regions, particularly the latter, were extremely poorly connected, since they were connected with only about 23 per cent of their neighbors. Steady improvement was seen through the economic rehabilitation and the First Five-Year Plan periods, when the proportion of non-connection at the higher level was exactly halved, and that at the lower level reduced from 76.67 to 51.67 per cent. The most drastic change occurred after the Second Five-Year Plan, in 1963, when only Yunnan and Tibet still lacked rail connection with Peking, and non-connection at the lower level was reduced to less than one-third. Closer examination further reveals that there appears to have been a conscious effort to spread improvement in railway connectivity evenly if only slowly, since 20 out of 24 incompletely connected provincial and autonomous regional capitals in 1949 having gained some degree of improvement by 1963, and since improvements clustered around two neighbors per province or autonomous region during any given period, with the exceptions of Hopei, where the national capital is located, Inner Mongolia, which has the maximum of eight neighbors, and Kansu, which improved its connection with 5 of its 6 previously unconnected neighbors. It should also be noted that by 1963, ten provinces and autonomous regions had in fact achieved complete connectivity, while another nine provinces and autonomous regions could be similarly considered if indirect connection was counted, reducing the number of provinces with less than complete connectivity with their neighbors to only seven.

In 1970, the higher level non-connection was again halved. In fact by then only Tibet remained unconnected with Peking. Improvement at the lower level connection however seemed to have slowed down somewhat throughout the Third and the Fourth Five-Year Plan periods. Yet it is perhaps less apparent that with the only exception of Tibet, and thus Tibet's neighboring administrative vertexes, all provinces and autonomous regions by 1970 had achieved complete connectivity either by direct or indirect rail, for, in addition to the twenty-three provinces and autonomous regions which had achieved complete connection by direct or indirect rail, four provinces and autonomous regions other than Tibet had direct and indirect

TABLE 27

RAILWAY CONNECTIVITY OF NATIONAL AND PROVINCIAL LEVEL CAPITALS*

Province	Capital	1949 (a)	1949 (b)	1952 (a)	1952 (b)
Anhui	Hofei	3/6+	1	3/6+	1
Chechiang	Hangchou	2/4	1	2/4	1
Chianghsi	Nanch'ang	5/6	1	3/6	1
Chiangsu	Nanking	0/3@	1	@	1
Chilin	Ch'angch'un	1/3	1	0/3@	1
Ch'inghai	Hsining	4/4	0	4/4	0
Fuchien	Fuchou	3/3	0	3/3	0
Heilungchiang	Harbin	1/2	1	0/2@	1
Honan	Chengchou	1/6	1	1/6+	1
Hopei	Shihshiachuang	2/5	1	0/5@	1
Hunan	Ch'angsha	6/6	0	3/6	1
Hupei	Wuhan	5/6	1	5/6	1
Kansu	Lanchou	6/6	0	6/6	0
Kuangtung	Canton	4/4	0	1/4	1
Kueichou	Kueiyang	4/4	0	4/4	0
Liaoning	Shenyang	1/3	1	0/3@	1
Shanhsi	Taiyuan	4/4	0	2/4+	1
Shantung	Chinan	0/4@	1	@	1
Shenhsi	Hsian	6/7	1	6/7	1
Ssuch'uan	Ch'engtu	8/8	0	8/8	0
Yunnan	K'unming	4/4	0	4/4	0
Autonomous Region					
Hsinchiang	Urumchi	3/3	0	3/3	0
Inner Mongolia	Huhehot	8/8	0	3/8	1
Kuanghsi	Nanning	4/4	0	2/4	1
Ninghsia	Yinch'uan	3/3	0	3/3	0
Tibet	Lhasa	4/4	0	4/4	0
Total not connected		92/120	14/26	70/120	9/26
Percentage not connected		76.67	53.85	58.33	34.62

*According to pre-1969 Administrative division, i.e., when the Inner Mongolian Autonomous Region had not been reduced in size. Nanking is exceptionally regarded as connected with Puk'ou across the Yangtzu river even before the completion of the bridge in 1968.

Notes:

(a) Number of neighboring provincial capitals (vertexes) not directly connected by railway over total number of neighboring capitals.

(b) Direct or indirect connection with Peking=1, non-connection=0. Total shows number of provincial level capitals not connected.

@ Complete connectivity with neighbors achieved.

+ Indirect connection with all neighboring capitals (complete connectivity) possible.

Direct and/or indirect connection with all neighboring capitals except Lhasa (Tibet).

TABLE 27—Continued

1957		1963		1970		1975	
(a)	(b)	(a)	(b)	(a)	(b)	(a)	(b)
3/6+	1	3/6+	1	3/6+	1	3/6+	1
2/4	1	1/4+	1	1/4+	1	1/4+	1
3/6	1	2/6+	1	2/6+	1	2/6+	1
@	1	@	1	@	1	@	1
@	1	@	1	@	1	@	1
4/4	0	2/4#	1	2/4#	1	2/4#	1
3/3	0	0/3@	1	@	1	@	1
@	1	@	1	@	1	@	1
1/6+	1	0/6@	1	@	1	@	1
@	1	@	1	@	1	@	1
1/6	1	1/6+	1	1/6+	1	1/6+	1
4/6	1	4/6+	1	3/6+	1	2/6+	1
5/6	1	1/6+	1	1/6+	1	1/6+	1
1/4	1	0/4@	1	@	1	@	1
2/4	1	2/4	1	0/4@	1	@	1
@	1	@	1	@	1	@	1
2/4+	1	0/4	1	@	1	@	1
@	1	@	1	@	1	@	1
5/7	1	3/7+	1	2/7+	1	2/7+	1
8/8	0	6/8	1	4/8#	1	3/8#	1
4/4	0	4/4	0	2/4#	1	2/4#	1
3/3	0	1/3#	1	1/3#	1	1/3#	1
3/8	1	2/8+	1	2/8+	1	2/8+	1
1/4	1	1/4	1	1/4+	1	1/4+	1
3/3	0	1/3+	1	1/3+	1	1/3+	1
4/4	0	4/4	0	4/4	0	4/4	0
62/120	7/26	38/120	2/26	30/120	1/26	28/120	1/26
51.67	26.92	31.67	7.69	25.00	3.85	23.33	3.85

connections with all other neighbors except Tibet. This situation was only marginally modified by 1975.

It seems clear therefore that the "dispersing" railway network of China and its "shifting" accessibility surface had to serve the objective of the organization of political areas; the meagre resources and progress must be maximized to satisfy the needs of different political nodes in the hierarchical system. To the extent that duplication and redundancy appears to have been avoided or at least minimized in achieving the objective, an efficient organization of political areas with the given network is apparent.

In Search of Development Goals

Railway Inputs and Development Priorities

The above analysis of the railway patterns of Communist China appears to indicate that there has been clear locational and development preferences in the evolving transportation system. To the extent that the railway has been the prominent circulation system in China during the period under study, and that the resultant structure has obviously been developed to facilitate areal organization and areal hierarchy, its locational patterns have much to contribute to the present day space-polity of the country. To explore further the functional relationship between political area and political idea in Communist China through transportation, it would be necessary to probe in the other direction, and to identify the objectives and priorities, which are manifestly determined by political ideology, in the development of the circulatory system.

The Data: For this purpose, it is proposed to employ regression techniques,[1] and the smallest spatial unit of analysis for which data can be assembled will be the province and autonomous region. For obvious reasons, the three municipalities of Peking, Shanghai, and Tientsin will not be treated as provincial level units, and for

---

[1]Multiple regression is an extension of the use of the bivariate correlation coefficient to multivariate analysis. It allows one to study the linear relationship between a set of dependent variables and a number of independent variables while taking into account the interrelationships among the independent variables. Analysis in this Section follows the computer package and procedure for multiple regression alalysis as contained in, Norman H. Nie, et al., Statistical Package for the Social Sciences (New York: McGraw-Hill, 1970), pp. 174-195, and its updating manuals.

want of consistency, all provincial and autonomous regional boundaries are those of the mid-1960's, so that boundary change of the Inner Mongolian Autonomous Region of 1969 has been ignored and post-1969 data have been appropriately transformed as far as possible.

The dependent variables for the analysis will be the amount of investment inputs crudely represented by the total railway route kilometrage added since 1949 and, as a control, the total operating route kilometrage achieved at the end of the selected periods. In order to take account of any effect of a pre-existing network on subsequent constructions, the first dependent variable will be transformed where necessary into a "shift-and-share index" of railway constructions using a formula developed by Fuchs[1] in an effort to produce comparative or refined results. The independent variables, the selection of which has been restricted by the availability of data, will include some 24 economic, physical, political, and social variables which are deemed to have had possible effects on the spatial distribution of railway investment input. These variables are listed in table 28, and their source and explanations are given in appendix E.

The analysis has been carried out for the years 1957, 1963, 1970, and 1975, roughly coinciding with the major planning periods.

---

[1]Fuchs uses the value-added data to calculate the gain or loss of manufacturing G_s, in the state s in the following manner:

$$G_s = Y_s - H_s, \text{ and } H_s = X_s \frac{Y}{X}$$

Substituting value-added by manufacturing with railway-added data in China, the above formula can be applied, where,

X_s = railway in km in the initial year (1949, etc.) in province or autonomous region s.
Y_s = railway in km in the selected year (1957, etc.) in province or autonomous region s.
X = railway in km in the initial year (1949, etc.) in China.
Y = railway in km in the selected year (1957, etc.) in China.

Thus, H_s is an abstract number representing the value of railway investment in km in province or autonomous region s that would exist if the province or autonomous region had grown at the national rate.

The difference between the actual value Y_s and H_s can then be converted into a percentage gain or loss, or _shift-and-share_ index, by the following:

$$\frac{100(Y_s - H_s)}{Y_s}$$

See Maurice Yeates, _An Introduction to Quantitative Analysis in Human Geography_ (New York: McGraw-Hill, 1974), pp. 100-122; and V. R. Fuchs, _Changes in the Location of Manufacturing in the United States since 1929_ (New Haven: Yale University Press, 1962), and "Statistical Explanations of the Relative Shift of Manufacturing among Regions of the United States," _Papers and Proceedings of the Regional Science Association_ 8, 1962, pp. 105-126.

TABLE 28

THE INDEPENDENT VARIABLES

Code	Description
CULTA	cultivated area
GRAOP	average grain output
GRAPP	average grain output per capita
GVIO	gross value of industrial output
GVIOP	gross value of industrial output per capita
OPRO	operating roads in kilometer
POCUL	population per unit of cultivated area
PODEN	population density
POP	population
TRANR	combined rankings by operating roads (OPRO) and navigable waterways (WAWAY), both in kilometers
URDEN	density of urban population
URPOP	urban population
WAWAY	navigable waterways in kilometer
AREA	total land area
DEGRE	degrees of longitude east
DEGRN	degrees of latitude north
HOSTI	per cent of hostile land
SHAPE	shape of province
ARDIS	air distance from Peking and Shanghai
CCPMB	Chinese Communist Party membership
LOCIN	index of "geopolitical" location
RSTAB	stability rank
PHAPC	per cent of farm households in Higher Agricultural Producers' Co-Ops
PMIN	per cent of non-Han population

Note: For sources and explanations, see appendix E.

Results for the first two periods are similar and thus 1949-1963 is considered one period. The years 1963-1970, essentially the period of the Cultural Revolution, exhibit features of confusion and "transition". Thus, for brevity and convenience, results will be presented only for two periods, namely, the pre-Cultural Revolution period, 1949-1963 and the post-Cultural Revolution period, 1963-1975 (which actually includes the period of the Cultural Revolution when there was less construction).

The Pre-Cultural Revolution Goals: Results of the multiple regression analysis show that (1) more than 99 per cent of the variation in the distribution of the total operating railways achieved in 1963 and (2) as much as 97 per cent of the variation in the distribution of the railways added between 1949 and 1963 can be accounted for by the chosen independent variables. Using a stepwise procedure[1], the most important variables explaining a high proportion of the variation can be identified.

Generally speaking, in terms of the "railways operating 1963", railway lines in that year were significantly concentrated in the eastern part of the country (DEGRE), though not particularly in the north (DEGRN) (table 29). Specifically, railways were found in areas of large urban population (URPOP), but away from areas of particularly high densities of population (PODEN); in areas of abundant cultivated land (CULTA), but away from those with high population densities per unit of cultivated land (POCUL); and, perhaps surprisingly, in areas of relatively good road transportation facilities (OPRO). More railway lines were also found in areas of high

---

[1]Step-wise regression is a powerful variation of multiple regression which provides a means of choosing independent variables which will provide the best prediction possible with the fewest independent variables. This computational method provides two pieces of information necessary to select the next variable to be brought into the equation. The first is the normalized regression-coeffieient value B that the prospective independent variable would have if it were brought into the equation on the next step. The significance of B is measured by the F statistic. If F is too small, there is little reason to add that independent variable to the prediction equation. The second piece of information is the pivot element which is known as the tolerance. If the tolerance approaches zero, then that variable is nearly a linear combination of variables already in the equation. A large tolerance indicates that a new "dimension" is being added to the prediction equation. The amount of additional variance explained by adding the new variable is the product of the normalized regression coefficient B squared and the tolerance. Thus, even if the prospective 3 is large, a small tolerance value will negate the value of that variable being added to the equation. Consequently, step-wise regression never brings a variable into the equation if the tolerance is below a specified minimum. See Norman H. Nie, et al., Statistical Package, p. 180.

TABLE 29

CORRELATION BETWEEN "RAILWAYS OPERATING 1963" AND SELECTED INDEPENDENT VARIABLES

Variable	B	Multiple R	R Square	Significance Level
DEGRN	-7.08628	0.56603	0.32039	–
DEGRE	99.38632	0.70982	0.50385	***
POCUL	-0.60388	0.79523	0.63239	–
OPRO	0.18322	0.87582	0.76706	***
PODEN	-5.40138	0.89823	0.80681	***
URPOP	0.10729	0.93596	0.87601	***
RSTAB	18.48189	0.94882	0.90025	**
CULTA	12.86307	0.95374	0.90961	***
PMIN	11.02166	0.96126	0.92401	**
Constant	-00041.25650			

*** 1% ** 5%

political stability (RSTAB), and with a larger proportion of minority nationalities (PMIN). When operating roads in kilometrage (OPRO) is replaced by combined rankings of operating roads and navigable waterways (TRANR), the results are practically the same. In other words, railways were found in areas generally with good transportation facilities, and particularly with good road trnasport rather than navigable waterways.

In summary the regional distribution of the railroad network in 1963 was highly correlated with eastward position in the territory of China, with a high urban population, with a large amount of cultivated land, and with a well-developed road network. Eastward location reflected both the greater development of the east compared to the west in China and location closer to the original nodes of railroad construction inward and landward from the coast. A high urban population, of course, needed railroads for transport of supplies and food to the cities and of goods from the cities. A large amount of cultivated land suggests the possibility of a regional agricultural surplus available for transport to other regions. A well-developed road network reflects the same favorable circumstances for substantial movement of goods and people that sustain railroad development also. The negative correlation with high density of population may reflect in part the subsistence nature of agriculture in areas with the highest population pressure on the land; such areas may have little surplus for trade or movement by railways out of the area to other regions

In terms of "railways added 1949-1963", however, the most significant variables are the shape (SHAPE) and the "geopolitical" location (LOCIN) of the area, which together account for over 56 per cent of the variation and are both highly significant (table 30).

TABLE 30

CORRELATION BETWEEN "RAILWAYS ADDED 1949-63" AND SELECTED INDEPENDENT VARIABLES[a]

Variable	B	Multiple R	R Square	Significance Level
SHAPE	227.76669	0.65978	0.43532	***
LOCIN	151.65889	0.75035	0.56302	***
HOSTI	2.12824	0.77673	0.60332	–
GRAOP	4.74600	0.80657	0.65055	***
DEGRE	38.22819	0.83375	0.69514	***
PODEN	-2.91164	0.86833	0.75400	***
GRAPP	-3.12469	0.90736	0.82329	***
CULTA	-8.59292	0.93191	0.86846	***
Constant	-2363.95920			

*** 1% ** 5%

[a]Excluding Tibet for lack of data

Though the amount of hostile land (HOSTI) is statistically not significant, it is still the third variable to enter into the question (table 30). Railroad construction in this period was definitely away from highly developed areas as revealed by negative correlations with population density (PODEN), average grain output per capita (GRAPP), and cultivated area (CULTA), as shown by table 30. On a simple regression basis (not shown in table 30) average grain output (GRAOP) and eastward location (DEGRE) also were negatively correlated with the dependent variable. New construction was away from areas of good transportation (TRANR), especially from the navigable waterways (WAWAY), though neither of these variables were among the most significant eight in table 30.

To separate out the effect of a pre-existing network from subsequent construction, a "shift-and-share index" of railway construction, 1949-1963, has been calculated with the same set of independent variables. The results show that both the total amount of variation explained (95 per cent), and the amount of variation accounted for by a combination of nine independent variables (88 per cent) remain high (table 31). Of the nine variables the proportion of hostile land (HOSTI) alone accounts for a very high proportion of the total variation, 46 per cent (able 31). Construction of railroads in this period was correlated also with high grain output per capita (GRAPP), characteristic of outlying areas. Negative correlations in table 31 reveal that construction of railroads 1949-1963 shifted away from the East and North (DEGRE, DEGRN), away from where waterway transportation was available (WAWAY), and away from areas of high population densities (PODEN).

TABLE 31

CORRELATION BETWEEN "SHIFT-AND-SHARE INDEX 1949-63" AND SELECTED INDEPENDENT VARIABLES

Variable	B	Multiple R	R Square	Significance Level
HOSTI	1.09492	0.67749	0.45899	***
GRAPP	0.13945	0.75121	0.56431	***
DEGRE	-0.65251	0.83906	0.70403	–
DEGRN	-1.71770	0.86130	0.74184	***
POCUL	0.12113	0.87979	0.77403	***
SHAPE	7.58915	0.90002	0.81004	***
WAWAY	-0.01653	0.91065	0.82928	***
PODEN	-0.24999	0.92359	0.85302	***
CCPMB	0.00557	o.93723	0.87841	**
Constant	22.21372			

*** 1% ** 5%

<u>The Post-Cultural Revolution Goals</u>: There may well have been two different sets of objectives and priorities in railway development in Communist China, one relating to the period 1949-1963, and the other to the period 1963-1975. For convenience, the former period has been referred to as the pre-Cultural Revolution period. The

latter period, in railway development, may be referred to as the post-Cultural Revolution period, since little construction took place before and during the Cultural Revolution, i.e., during the years 1963-1965 and 1966-1968. It would therefore be interesting to further the analysis by treating railway investment in the two periods ending in 1970 and 1975 as one, in the hope that the difference in the development strategies can be better focused.

Thus the "railways added" and the "shift-and-share index" dependent variables have again been modified to reflect only construction between 1963 and 1975, or rather during the post-Cultural Revolution period, thereby eliminating the effect of an earlier period when development strategies might have been different.

Table 32 shows that, in terms of railways added 1963-1975, unlike the pre-Cultural Revolution period, variables such as per cent of hostile land (HOSTI) and the index of geopolitical location (LOCIN) are no longer statistically significant. Of the first four variables which are statistically significant, three, cultivated area (CULTA), low population density (PODEN), and average grain output (GRAOP), are economic variables, and together they account for over half of the variation. Indeed, priorities of construction between 1963 and 1975 seem to have been given to areas with abundant cultivated land, high grain output, and relatively low population density, and at relatively short distances from Peking and Shanghai (ARDIS), with greater political stability. Generally, these are provinces with good economic, especially agricultural, potentials, such as Hupei, Hunan, Ssuch'uan, Anhui, Honan, Shenhsi, and Kueichou.

TABLE 32

CORRELATION BETWEEN "RAILWAY ADDED 1963-75" AND SELECTED INDEPENDENT VARIABLES

Variable	B	Multiple R	R Square	Significant Level
CULTA	8.17505	0.46751	0.21856	***
PODEN	-1.65536	0.67774	0.45933	***
GRAOP	2.02877	0.72992	0.53279	***
ARDIS	-0.30000	0.76246	0.58135	***
HOSTI	3.70205	0.77366	0.59856	–
CCPMB	-0.03087	0.78930	0.62299	–
LOCIN	21.21682	0.79953	0.63925	–
Constant	311.78985			

*** 1% ** 5%

Again this is more sharply focused by using the "shift-and-share index 1963-1975," as is shown in table 33. Here the only positively correlated variables are gross value of industrial output per capita (GIVOP), cultivated area (CULTA), and per cent of hostile land (HOSTI). The first two variables alone account for over 15 per cent of the total variation. Although the gross value of industrial output per capita (GVIOP) is not statistically significant, the fact that it is the first to enter the equation and accounts for over 9 per cent of the variation is very revealing. The remaining variables, which are all negatively correlated and all except Chinese Communist Party membership (CCPMB) statistically not significant,

TABLE 33

CORRELATION BETWEEN "SHIFT-AND-SHARE INDEX 1963-1975" AND SELECTED INDEPENDENT VARIABLES

Variable	B	Multiple R	R Square	Significance Level
GVIOP	0.00318	0.30623	0.09378	–
CULTA	0.08657	0.39281	0.15430	***
CCPMB	-0.00709	0.45400	0.20612	***
HOSTI	0.33783	0.49183	0.24189	–
SHAPE	-4.07242	0.53251	0.28357	–
LOCIN	-0.81666	0.56815	0.32229	–
DEGRE	-1.33164	0.58845	0.34628	–
ARDIS	-0.01935	0.60659	0.36795	–
DEGRN	-1.16835	0.63663	0.40530	–
Constant	187.04814			

*** 1%   ** 5%

strongly suggest that construction paid no attention to the Chinese Communist Party membership (CCPMB), the shape (SHAPE) or the geopolitical location (LOCIN) of the areas. Locationally, the shift in share index of railroad length was found close to the political core of Peking, and Shanghai, but not so much in the east (DEGRE) or north (DEGRN). The hostile environment (HOSTI) which had been a most important factor in the pre-Cultural Revolution period, is now not statistically significant though positively correlated.

<u>The Railway Network in 1975</u>: The achieved system in terms of railways operating in 1975 (table 34), the cumulation of all previous periods of railroad building, is highly correlated with the amount of cultivated land (CULTA), the gross value of industrial output per capita (GVIOP), the geopolitical location (LOCIN), and the population per unit of cultivated land (POCUL) of the provinces or autonomous regions. Locationally, these are in fact in the east (DEGRE) and not so densely populated areas (PODEN), namely, Heilungchiang, Inner Mongolia, Chilin, Liaoning, Hopei, Ssuch'uan, Kansu, and Honan.

TABLE 34

CORRELATION BETWEEN "RAILWAYS OPERATING 1975" AND SELECTED INDEPENDENT VARIABLES

Variable	R	Multiple R	R Square	Significance Level
CULTA	42.42060	0.60188	0.36226	***
PODEN	-6.29579	0.79478	0.63167	***
GVIOP	1.36307	0.91369	0.83484	***
DEGRE	55.46207	0.95128	0.90494	***
LOCIN	97.51627	0.95692	0.91570	***
POCUL	0.67481	0.95862	0.91895	**
SHAPE	59.35650	0.96059	0.92273	—
GRAPP	-1.01019	0.96318	0.92772	—
CCPMB	-0.05222	0.96536	0.93191	—
Constant	-5719.92509			

*** 1%   ** 5%

The most meaningful variables in the equation, however, are gross value of industrial output per capita (GVIOP) and index of geopolitical location (LOCIN). The former was not at all significant during the earlier pre-Cultural Revolution period 1949-1963 yet it now accounts for more than 20 per cent of the variation and is statistically highly significant. The latter, though important in terms of railways added during the pre-Cultural Revolution period, was never a factor the achieved system as a whole was able to cater to. Insofar as the achieved system is now correlated with the agricultural potential and output, industrial output, as well as the geopolitical needs of Communist China, it can be argued that the railway network as a circulatory system has clearly been developed to even better suit the physical, economic, and political space of the country.

Communist Development Strategy and Dialectics

It therefore seems clear that railway development strategies in China up to 1963 paid relatively little attention to economic considerations.[1] Rather, with the exception of Tibet, developments seem to have been located in environmentally "hostile" and geopolitically sensitive areas, namely, in Hsinchiang, Yunnan, Kuanghsi, and Fuchien; or in economically underdeveloped areas where there are low grain output per capita, low population density, and small cultivated area, such as Inner Mongolia, Ninghsia, Kansu, Ch'inghai, Hsinchiang, Yunnan, Kueichou, Kuanghsi, and Fuchien. In other words, with the exception of Fuchien and Kuanghsi, which lie in the third step of the physical space of China, all of these province and autonomous regions are found on the second and first steps, and in terms of economic development they constitute the "extra-ecumenical" areas or the "periphery" of China.

Thus identified, the periphery or the "extra-ecumenical" areas of China, compared to the "core" or the "ecumenical" areas, have been consistently favoured in terms of post-1949 railway development. In 1949, only about 12 percent of the operating railway route kilometrage was found in the "extra-ecumenical" areas, which, through the first two five-year plans, was rapidly increased to some 33 per cent in 1963 (fig. 26). The "extra-ecumenical" areas continued to improve its share of "railway operating" to more than 36 per cent in 1970 and only by then did it seem to stabilise.

Though a similar picture is depicted in the case of gross value of industrial output (fig. 27), the change was within an extremely narrow range, and by 1973, the "ecumenical" areas still accounted for over 96 per cent of the total gross value of industrial output. When it is recalled that this extreme imbalance is common to patterns of distribution of population, land resources, etc., it will be realized that railway development not only drained heavily resources from the "core" or "ecumenical" areas, but represented a heroic attempt at inter-regional balance and spatial integration.

---

[1]See also Alan P. L. Liu, Communications and National Integration in Communist China, p. 13.

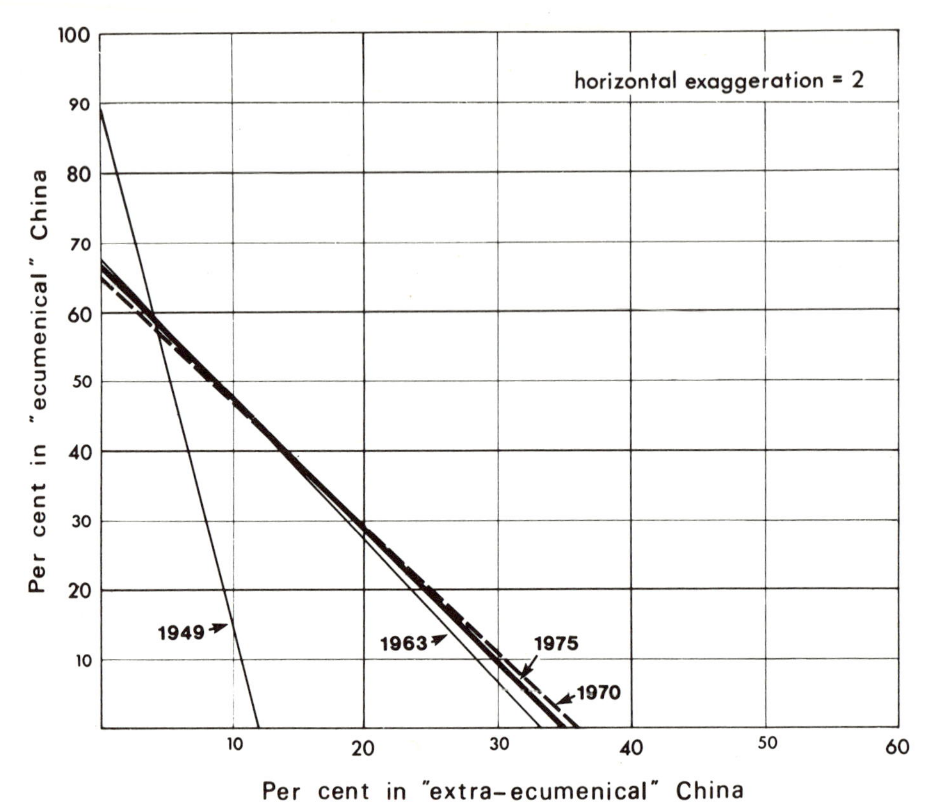

Fig. 26. Distribution of operating railways in "ecumenical" and "extra-ecumenical" China.

Summary of Findings

The development of the most important sinew of the Chinese space-polity, the railway system, was for the first time in the history of railway development in China completely free from practically all external pressures and interferences. Neither was there much difference internally over the role and functions of the railways, for communist ideology provided the unifying basis for national development planning

Railway development before 1963 generally conformed to Mao's dialectical method of solution of these contradictions over the Chinese territorial space, i.e., through the development of the "opposite" or the "extra-ecumenical" areas of China, although such as effort presumably involved huge regional transfer of resources. In short, that the People's Republic of China emphasized territorial

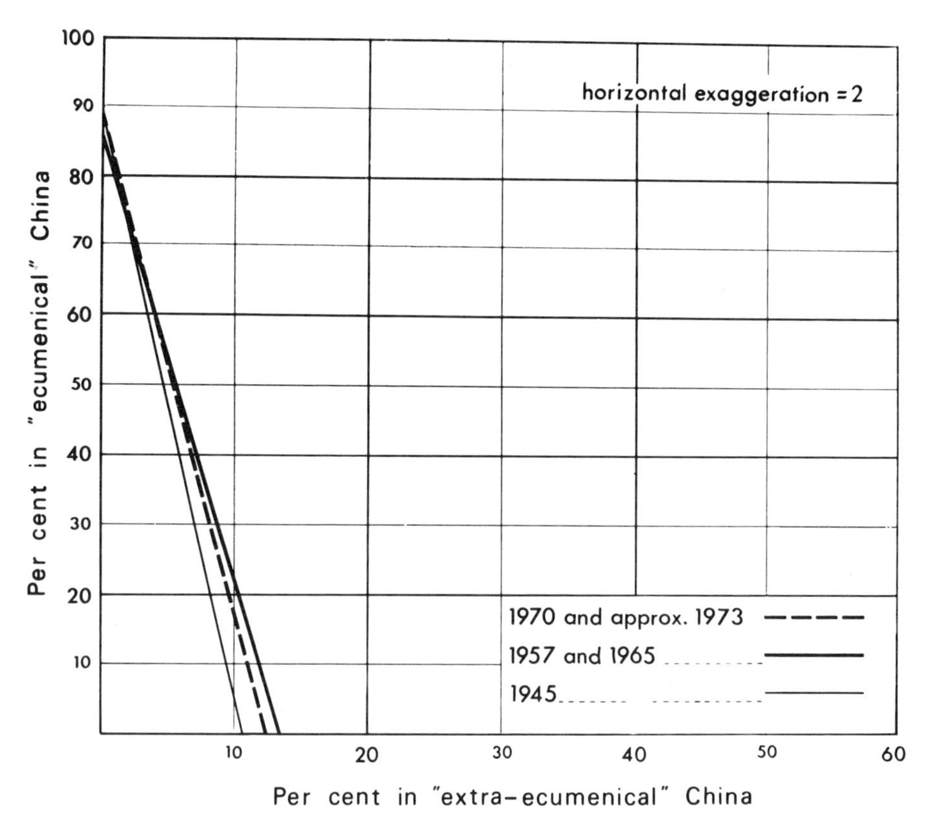

Fig. 27. Distribution of gross value of industrial output in "ecumenical" and "extra-ecumenical" China.

integration in her approach to railway development before 1963 was as much a reaction to foreign encroachment as a conscious effort to resolve the contradiction of regional imbalance in political organization that she had inherited.

Constrained not just by political considerations, but also by physical factors, the direction of spatial integration in China has been generally towards the continental frontiers which are now characterised as the peripheral or "extra-ecumenical" areas and, at the same time, towards the more hostile environment of the second step, where Chinese peasants and migrants throughout history have attempted to establish themselves. Seen in this light, a successful integration of the second and ultimately the first steps is tantamount to a resolution of the most important contradiction that exists over the political as well as the physical space of China.

Even so, the dialectical method was not applied statically. For after 1970, or more accurately since the Cultural Revolution, when the pattern of the expanding railway system was sufficiently dispersed, the dynamic development of the state necessitated a shift of emphasis towards the core. To the extent that the achieved pattern in 1975 is capable of catering to the economic, physical, and geopolitical needs of the country, it is clear that the transportation system faithfully reflects the ideologies and therefore national goals of Communist China.

The Paot'ou-Lanchou Railway passing through the Tengri Desert

Sections of the Paochi-Ch'engtu Railway in mountainous Shenshi

The Yingt'an-Hsiamen Railway

The Ch'engtu-K'unming Railway overcoming steep gradient and difficult terrains.

The laying of 1000 m long rails on the Peking-Tientsin Line

Modern transportation facilities at the service of China's minority nationalities, the Yuanmou station on the Ch'engtu-K'unming Railway

Railway builders participate in conservation work along the Chiaotso-Chihcheng Railway

A train runs through the tracks of the Southern Hsinchiang Railway, while construction obviously continues.

PART THREE

CONCLUSION

CHAPTER VI

IDEOLOGY, TRANSPORTATION, AND SPACE-POLITY

Railway development in China has been a symbiosis of an evolving *raison d'etre* on the one hand and a transforming space-polity on the other. During the same period, the state itself too has progressed from a quasi-feudal to a modern and then a full-grown communist state. Pre-modern China, like other pre-modern states, was only vaguely bounded, with a space-polity which exhibited distance-decay of authority from the center of power. Under the Ch'ing dynasty the dominant ideology was the maintenance of minority rule, which was found unsuited to the Han majority in general and to the multi-ethnic nature of China in particular. This was the time when modern nation-states emerged and when strenuous efforts were made to expand and demarcate national territories. The weakness of China, particularly since the beginning of the Nineteenth Century, coupled with her peculiar *raison d'etre* and the nature of her space-polity, facilitated foreign encroachment.

If the Ch'ing dynasty clung to an outmoded ideology, the situation under the Republic was no better, for Sun Yat-sen's *san-min-chu-i* ran counter to the basic values of democracy and capitalism,[1] and its dominant ideology of nationalism, which rallied the population for the overthrow of the Manchu Empire, was at best a negative one from the standpoint of frontier affairs in a multi-ethnic nation such as China.[2] Yet the national aspiration for modernization that started more than a century ago, which had once swayed from anti-foreignism to complete Westernization, had at least found new

---

[1]See discussion under chapter 3, section 1.

[2]George Moseley, "The Frontier Regions in China's Recent International Politics," pp. 299-329. Paul Linebarger once remarked, "From Han (times) down to the present, the border areas have been centers of Chinese strategic thinking," as quoted in P. M. Roxby, "China as an Entity: The Comparison with Europe," *Geography*, 19 (March 1934): 1-20.

confidence in the rising sense of modern nationalism. On the other hand, the legacy of the semi-colonial status of the Ch'ing dynasty led to a divided space-polity and eventually the Sino-Japanese War, and the rising tide of nationalism which naturally joined forces with communism to the civil war that followed, preventing systematic and purposeful development of the circulatory system, particularly railway transportation.

In the People's Republic of China, Mao Tse-tung was instrumental in the visualization of the communist society and in his interpretation and application in China of Marxism-Leninism or dialectical materialism. Indeed, Mao Tse-tung thought, as it is known, or dialectical materialism expressed in Chinese language and form,[1] represented not only a new political ideology but a nationally accepted iconography developed in harmony with both international politics and national aspiration,[2] at least until his death. The national goals or aspirations for modernization and for national integrity of more than a century found expression in Mao's dialectical analysis of the Chinese environs. His analysis highlighted the many spatial contradictions that national development, as well as the development of the circulatory system, must attempt to overcome. The railway, therefore, was not only a means of territorial integration and national defence, but a vital sinew for areal organization. It must also contribute to overcoming traditional physical barriers and to strengthening geopolitically sensitive areas. In short, it must be an effective system of areal administration in time of peace, and of military mobilization in time of war. The achieved system appears to fulfil all these goals.

If the three regimes have had divergent experiences and fortunes in the development of the circulatory system, governed by the different ideologies on the one hand and reflected in the different areal patterns on the other, they at least share some commonalities. First, behind all questions of national development in China, there has been the assumption that traditionally the state rather than the individual should have the major role in providing and

---

[1]Steve S. K. Chin, Mao Tse-tung, and C. K. Leung, "A Review of the Thought of Mao Tse-tung."

[2]For, as Ladis K. Kristof states, "the iconography of the fatherland is ever being adjusted to harmonize with evolving aspirations." See Ladis K. D. Kristof, "The State-Idea, and National Idea and the Image of the Fatherland," Orbis 11 (September 1967): 238-255.

enforcing the answers.[1] This has been most apparent in the thinking of Li Hung-chang, Sun Yat-sen, and Mao Tse-tung, the three national leaders during various periods of railway development in China. Second, in the development of a national system of transportation, there has been a consistent attitude towards its political and military role. Both of these commonalities in fact readily find parallels in Chinese history. Indeed, from the time when the First Emperor of Ch'in, for effective unification of the new Empire, standardized among other things the gauges of all wheeled carts, chariots, etc. the complicated but well-planned system of (horse) roads, stops, and post-stations, of the Ch'in dynasty, together with its great military significance had been well recognized.[2] Even the canal system, part of the transportation network of old China, conformed more to the political needs than to commercial desirability. The first important canals were dug in the Han dynasty basically for improving mobility from and around the capital, Ch'angan, though they were also useful for irrigation. The most famous canal, the Grand Canal, dug in Sui times, was to facilitate the movement of tributary offerings from the then rich South, on which the central government depended for its survival.[3]

As a modern institution, the railway system is no exception. Specifically, it has been repeatedly regarded as an important means for territorial integration and national defence. The railway was first reluctantly accepted by the Imperial Ch'ing government primarily as a means of defence and subsequently used, though in vain, as a means of reintegrating the Northeast and of strengthening China Proper. Similarly, the Republic of China first saw through Sun Yat-sen national salvation by means of railway construction, and later employed practically the same strategies as her predecessor of a military railway program.

What makes the Communist approach more distinctive from that of the Ch'ing and the Republic is twofold. First, although the political role of the railway has been recognized and territorial integration with a view to the strengthening of national defence was clearly an objective in railway-building, the distribution of the system was specifically designed to eliminate gross regional contradictions, whether physical, political, economic, or even social,

---

[1]Theodore Herman, "Group Values toward the National Space: the Case of China" pp. 164-182.

[2]Pai Shou-i, Chung-kuo chiao-tung shih [A history of China's communications] (Taipei: Commercial Press, 1965), p. 82.

[3]Ibid., pp. 86 and 109.

that existed in China's territorial space. There has also been a deliberate attempt at harnessing the second step of the physical environment to breathe life into the Lebensraum of China. Second, there is an additional emphasis on the administratitive function of the railway, for its construction obviously contributed to a sound areal hierarchy and thus to the integrity of the Chinese space-polity. Yet, seen from the standpoint that an effective areal hierarchy is a prerequisite to territorial integration and therefore national defence, it is only an added element in essentially the same set of spatial strategies adopted by the present government's predecessors.

This leads to the obvious conclusion that different goals can be attained with identical strategies, but that development strategies or even political decisions themselves are ineffective if not framed in an accepted raison d'etre, or can even be counterproductive in the absence of an appropriate political ideology. A close and harmonious relationship between "iconography", "circulation", and "political area" is clearly fundamental to the coherence and viability of a state.

APPENDICES

APPENDIX A

FOREIGN RAILWAY CONCESSIONS IN CHINA 1895-1911 & 1912-1914

Year	Line	Length (km)	Location	Type of Rights
			BRITAIN	
1895-1911				
1898	Canton-Kowloon	171.77	Kuangtung	L.M.O.
1898	Pei-Ning	1,388.55	Hopei, Liaoning	L.M.O.
1898	Peking-Shanghai	327.13	Chiangsu	L.M.O.
1898	Shanghai-Hangchou-Ningpo	186.53	Chiangsu, Chechiang	L.M.O.
1898	Pu-Hsin	563.15	Anhui, Honan	L.
1898	Tao-Ch'ing	165.44	Honan, Shanhsi	C.
1899	Tientsin-Puk'ou	1,083.78	Hopei, Shantung, Anhui, Chiangsu	L.
1904	Tibet area	—	Tibet	C. priority
1907	Hsin-Fa	80.45	Liaoning	L.
1909	Chin-Ai	1,500.00	northeastern provinces	L.
1911	Yueh-Han-Ch'uan	1,800.00	Kuangtung, Hunan, Hupei, Ssuch'uan	L.
1912-1914				
1913	Hsia-Hsing	1,053.89	Hunan, Hupei, Kuanghsi	L.
1913	Yun-Ta	550.00	Yunnan	L.
1913	Ning-Hsiang	1,600.00	Chiangsu, Anhui, Chiangshi, Hunan	L.
1914	Kuang-K'an	800.00	Chianghsi, Kuangtung	L.
1914	Kuang-Ch'ao	450.00	Kuangtung	L.
1914	Ch'ao-Fu	450.00	Kuangtung, Fuchien	L.
1914	Fu-K'an	450.00	Chianghsi, Fuchien	L.

APPENDIX A--Continued

FRANCE

1895-1911				
1896	Lung-Chou	80.45	Kueichou	C.
1897	Ping-Han	1,332.93	Hopei, Honan, Hupei	L.
1898	Tien-Yueh	470.00	Yunnan	C.
1898	Chih-An	96.54	Kuangtung	C.
1898	Nan-Ning	193.08	Kuangtung, Kueichou	C.
1903	Pien-Lo[a]	185.13	Honan	L.M.O.
1911	Yueh-Han-Ch'uan	1,800.00	Kuangtung, Hunan, Hupei Ssuch'uan	L.
1912-1914				
1912	Lung-Hai	1,808.00	Chiangsu, Honan, Shenhsi	L.M.O.
1913	Tung-Ch'eng	1,600.00	Ssuch'uan, Shenhsi, Shanhsi	L.M.O.
1914	Ch'in-Yu	2,043.43	Kuangtung, Kuanghsi, Yunnan, Kueichou, Ssuch'uan	L.M.O.
1914	Hsu-Ch'eng	280.00	Ssuch'uan	L.
1914	Kuanghsi area	—	Kuanghsi	L.

RUSSIA

1895-1911				
1896	Chinese Eastern	1,721.00	Chilin, Heilungchiang	Joint venture
1897	Ping-Han[a]	1,332.93	Hopei, Honan, Hupei	L.
1898	South Manchuria	844.80	Chilin, Liaoning	Joint venture
1898	Cheng-T'ai	249.95	Hopei, Shanhsi	L.M.O.
1899	north of Peking	—	provinces north of Peking	L. priority
1902	Chi-Ch'ang	127.70	Chilin	C.
1912-1914				
1913	Tao-Ang	224.20	Liaoning, Heilungchiang	C.
1914	Outer Mongolia Area	—	Outer Mongolia	L. priority

GERMANY

1895-1911				
1898	Chiao-Chi	445.99	Shantung	C.
1899	Chiao-I-Chi	—	Shantung	C.
1899	Tientsin-Puk'ou	1,083.78	Hopei, Shantung, Anhui Chiangsu	L.
1902	Kai-Chi	500.00	Honan, Shantung	L. priority
1902	Ching-Teh	390.00	Hopei, Shantung	L. priority
1911	Yueh-Han-Ch'uan	1,800.00	Kuangtung, Hunan, Hupei, Ssuch'uan	L.

		GERMANY		
1911-1914				
1913	Kao-Han	337.98	Shantung, Chiangsu	L.
1913	Shun-Chi	193.08	Shantung, Hopei	L.
1913	Yen-Wei	229.00	Shantung	L. priority
		JAPAN		
1895-1911				
1905	South Manchuria	844.80	Liaoning, Chilin	Joint venture
1905	An-Shen	260.20	Liaoning	" "
1906	Nan-Hsin	128.35	Chianghsi	L.
1907	Hsin-Shen	59.77	Liaoning	L.M.O.
1907	Chilin-Ch'angch'un	127.70	Chilin	L.M.O.
1909	Chilin-Huining	320.17	Chilin	L. priority
1912-1914				
1913	Ssu-Tao	312.11	Liaoning	L.M.O.
1913	Ch'ang-Tao	240.00	Liaoning, Chilin	L. priority
1913	Kuan-Hai	120.00	Liaoning	L. "
1913	Hai-Chi	250.07	Chilin	L. "
1913	Tao-Je	756.23	Liaoning, Jeho	L. "
1914	Ch'i-Yu	24.00	Liaoning	Joint venture

[a]Although concessions for these two lines were given to a Belgian syndicate, which was in fact directly controlled by France and Russia, they are listed under these countries rather than Belgium.

Abbreviations: L. = Loan
M. = Management
O. = Operation
C. = Construction

SOURCE: Compiled and translated from Ch'en Hui, Chung-kuo t'ieh-lu wen-t'i, pp. 30-31 and 39-40.

APPENDIX B

RAILWAY TRANSPORTATION DATA

Length of identified railways in operation by railway lines and province

Line	Terminals		Length (km) in Operation at End of Indicated Year					
	From	To	1949[a]	1952[a]	1957[a]	1963[a]	1970	1975
	PROVINCE							
	ANHUI							
Ch'ing-Pu	T'aoshanchi[b]	Tungko	294	294	294	294	294	294
Ning-Kan	Tz'uhu	Wuhu	10	10	50	50	50	50
Huai-Nan	Pangfou	Yuch'ik'ou	131	248	248	248	248	248
	Pakungshan	Shuichiahu	27	27	46	46	46	46
	Tat'ung	Tienchiaan	—	—	6	6	6	6
	Wuhu	Fenghsiangk'uo	—	—	—	53	53	53
	Fenghsiangk'uo[c,d]	Tungling	—	—	—	—	19	54
Lung-Hai	T'angshan[b]	Hoisai	61	68	68	68	68	68
	Fulichi	Huaich'i	—	—	—	31	31	31
	Huaich'i[d,e]	Fuyang	—	—	—	—	154	154
	Huaich'i[d]		—	—	—	—	66	66
		Total	523	647	712	796	1035	1070
	CHECHIANG							
Hu-Hang-Yung	Fengching[f]	Ningpo	140	140	294	294	294	294
	Nanhsingch'iao	Hangchou	—	—	—	3	3	3
	Hangchou[g]	Ch'anghsing	—	—	—	—	—	90

Che-Kan	Hsiaoshan	Hsint'angpien	302	302	302	302	302	302
	Chinhua	Hsinanchiang	–	23	23	69	69	69
		Total	442	465	619	668	668	758
	CHIANGHSI							
Che-Kan	Hsint'angpien[h]	P'inghsiang	570	570	570	570	570	570
	Changchiashan[i]	Shangtang	–	–	–	–	51	51
	Ch'uanchiang	Kaok'eng	–	7	7	6	6	6
Nan-Hsun	Nanch'ang[h]	Chiuchiang	135	135	135	135	135	135
Ying-Hsia	Yingt'an	Tzuch'i	–	–	73	73	73	73
	T'ankang[c,d]	Kungch'i	–	–	–	–	125	125
		Total	705	712	785	784	960	960
	CHIANGSU							
Ch'ing-Pu	Hanchuang[b]	T'aoshanchi	66	66	66	66	66	66
	Tungko[b]	P'uk'ou	18	18	23	23	23	23
	P'uk'ou	Nanking	–	–	–	–	20	20
	Hsuchou E.	Hsuchou N.	–	–	6	6	6	6
Te-Shih	Liuch'uan	Chiawang	–	–	16	16	16	16
Hu-Ning	Shanghai	Nanking	312	312	312	311	311	311
	Shanghai	Wusung	16	16	16	15	15	15
	Shanghai	Wentsaopin	–	–	–	14	14	14
	Shanghai	Minhang	–	–	–	42	42	42
Hu-Hang-Yung	Shanghai	Fengching	70	70	70	70	70	70
	Jihhuikang	Hsinlunghua	–	4	4	4	4	4
Ning-Kan	Nanking	Tz'uhu	81	81	81	81	81	81
Lung-Hai	Lienyunkang[b]	Hoisai	238	238	238	238	238	238
		Total	801	805	832	886	906	906

APPENDIX B--Continued

Line	Terminals		Length (km) in Operation at End of Indicated Year					
	From	To	1949[a]	1952[a]	1957[a]	1963[a]	1970	1975
	PROVINCE							
	CHILIN							
Shen-Chi	Chaoshih[b]	Chaoyangchen	102	102	102	102	102	102
	Ch'aoyangchen[d]	Shansongang	—	—	—	—	31	31
Shen-Pin	Ssup'ing[b]	Hsichia	296	296	296	296	296	296
	Taolaichao	Yushu	—	—	56	56	56	56
Shen-Chi	Ch'aoyangchen	Chilin	184	182	183	183	183	183
	Yentungshan[d]	Huatien	—	—	—	—	—	74
	Lungt'anshan	Tafengman	—	23	23	23	23	23
	Chilin	North Chilin	—	—	12	12	12	12
	Shulan[b,d]	Chilin	—	30	30	30	88	88
Ssu-Chi	Ssup'ing	Chian	401	400	400	400	400	400
	T'unghua	Hsint'unghua	—	—	—	4	4	4
	Hunchiang	Wank'ou	—	18	18	48	48	48
	Yayuan	Talintzu	113	113	113	113	113	113
	Wank'ou[c,d]	Lushuiho	—	—	—	—	62	117
Ch'ang-A	Kouch'iench'i	K'okenmiao	119	184	238	238	238	238
	Kokenmiao*	Aerhshan	—	—	—	—	—	—
	Ch'angch'un[e]	Kuoch'iench'i	—	—	—	—	—	166
	Taan[d]	Tungliao	—	—	—	—	176	176
Ch'ang-Tu	Ch'angch'un	T'umen	528	527	529	529	529	529
	Hsinchan	Lafa	—	9	8	8	8	8

	Lungching	Holung	51	51	51	51	51	51
	Chiangpei	Chinchu	—	19	19	19	19	19
	Ch'iaoho	Naitzushan	—	10	10	10	10	10
	Ch'aoyangch'uan	Kaishants'un	58	58	58	58	58	58
Mu-T'u	Lutao	T'umen	146	146	146	146	146	146
	Wangch'ing	Hsiaowangch'ing	—	9	9	9	9	9
La-Pin	Hsiaokuchia	Shuichuliu	97	97	97	97	97	97
Ta-Cheng	Aerhhsiang*	Chengchiats'un	—	—	—	—	—	—
P'ing-Ch'i	Sanchangk'ou[b]	Paich'engtzu	—	289	289	289	289	289
	Paich'engtzu[b]	T'ant'u	—	78	78	78	78	78
		Total	2095	2641	2765	2799	3126	3421
	CH'INGHAI							
Lan-Ch'ing	Shuich'ewan	Hsining	—	—	—	122	122	122
Hsi-Hai	Hsining	Haiyen	—	—	—	100	100	100
	Hsining[c]	Tat'ung	—	—	—	—	42	42
		Total	—	—	—	222	264	264
	FUCHIEN							
Ying-Hsia	Tzuch'i	Laichou	—	—	212	212	212	212
	Laichou	Hsiamen	—	—	409	409	409	409
Lai-Fu	Laichou	Fuchou	—	—	—	194	194	194
	Kuok'eng	Lungch'i	—	—	—	11	11	11
	Changp'ing	Lungyen	—	—	—	58	58	58
		Total	—	—	621	884	884	884

APPENDIX B--<u>Continued</u>

Line	Terminals		Length (km) in Operation at End of Indicated Year					
	From	To	1949[a]	1952[a]	1957[a]	1963[a]	1970	1975
	PROVINCE HEILUNGCHIANG							
Mu-T'u	Mutanchiang	Lutao	102	102	102	102	102	102
La-Pin	Shuichuliu	Hsiangfang	163	163	163	163	163	163
Pin-Pei	Harbin	Peian	326	333	333	333	333	333
	Peian[d]	Lungchen	—	—	—	—	55	55
	Harbin[h]	Sank'oshu	11	11	11	11	11	11
	Linhai[e]	Pishui	—	—	—	—	119	119
Ch'i-Pei	Ch'ich'ihaerh	Peian	—	231	231	231	231	231
	Fuyu	Nenchiang	181	182	180	180	180	180
	Nenchiang[d,*]	Jagdaqi	—	—	—	—	—	—
	Jagdaqi[d,e,*]	Linhai	—	—	—	—	—	—
	Linhai	Changling	—	—	—	—	230	230
	Changling	Kulien	—	—	—	—	—	130
Sui-Chia	Suihua	Lienchiangk'ou	369	369	369	369	369	369
	East Chiamussu	Shuangyashan	—	—	72	72	72	72
Shen-Pin	Hsichia	Harbin	62	62	62	62	62	62
P'ing-Ch'i	T'antu[b]	Ch'ich'ihaerh	—	140	140	140	140	140
	Hungch'iying	Yushuts'un	—	—	3	3	3	3
Sui-Man (n.)	Nientzushan*	Manchouli	—	—	—	—	—	—
Ya-Lin	Yak'oshih*	K'utuerh	—	—	—	—	—	—
	K'utuerh*	Kanho	—	—	—	—	—	—

	Kanho*	Jagdaqi	–	–	–	–	–	–
	Chaochung*	Mordaga	–	–	–	–	–	–
	It'uliho*	Kenho	–	–	–	–	–	–
	Kenho*	Mankuei	–	–	–	–	–	–
Chia-Mu	Mutanchiang	Hokang	392	392	392	392	392	392
	Poli[h]	Chitaiho	–	–	–	31	31	31
Chia-Lien	Chiamussu	Lienchiangk'ou	–	–	6	6	6	6
Lin-Hu	Link'ou	Hut'ou	172	171	171	336	–	–
	Link'ou[d]	Tungfanghung	–	–	–	–	331	331
	Hsichihsi	Hsiach'engtzu	104	104	103	103	103	103
	Chihsi	Hengshan	12	12	12	12	12	12
Sui-Man	Suifenho	Nientzushan	903	903	903	903	903	903
	Kouk'ou*	125 kilometer	–	–	–	–	–	–
	Yapuli	Loushan	–	70	70	70	–	–
	Naot'ou	Ch'angting	–	–	–	42	42	42
	Saerht'u[d]	Taan	–	–	–	–	154	154
Nan-Lin	Nani	Ich'un	105	105	104	104	104	104
	Ich'un[h]	Hsinch'ing	–	–	108	108	108	108
	Ich'un	Tsuiluan	–	–	21	21	21	21
	Hsinch'ing[d,e]	Wuyiling	–	–	–	23	43	54
		Total	2902	3350	3556	3817	4320	4461
	HONAN							
Ching-Han	Tz'uhsien	Wushengkuan	562	562	562	562	562	562
	T'angyin	Haopi	–	–	19	19	19	19
	Hsinhsiang	Chiaotso	64	64	62	62	62	62
	Chiaotso	Poai	–	–	–	–	32	32

APPENDIX B--Continued

Line	Terminals		Length (km) in Operation at End of Indicated Year					
	From	To	1949[a]	1952[a]	1957[a]	1963[a]	1970	1975
	PROVINCE							
	HONAN							
	Hsinhsiang[d]	Fengch'iu	—	—	—	—	—	66
	Wulifou	Chengchou	—	—	6	6	6	6
	Mengmiao	Shenlou	—	—	61	61	61	61
	Shenlou[d]	Baofeng	—	—	—	—	35	35
Lung-Hai	Tangshan[h]	T'ungkuan	642	642	642	642	642	642
	Huihsing	Sanmenhsia	—	—	14	14	14	14
	Loyang	Iyang	—	—	—	22	22	22
	Anyang	Lichen	—	—	—	32	32	32
	Hsiangfan[d]	Chiaotso	—	—	—	—	506	506
		Total	1263	1268	1366	1420	1993	2059
	HOPEI							
Ching-Han	Peking	Tz'uhsien	481	481	481	481	481	481
	Fengtai[d]	Laiyuan	—	—	—	—	154	194
	Chouk'outien	Liuliho	—	—	15	15	15	15
	Paoting	Paotingnan	—	—	7	7	7	7
	Hantan	Hots'un	—	—	50	50	50	50
	Hots'un[h]	Fengfeng	—	—	—	29	29	29
	Mat'ou	Fengfeng	—	—	19	19	19	19
	K'uangshan	Talien	—	—	35	35	35	35

	Paoting[e]	Mancheng	—	—	—	—	—	20
	T'unghsien	Kuyeh	—	—	—	—	—	186
Ching-Pao	Hsichihmen[h]	Hsiwanp'u	238	238	238	238	238	238
	Changchiak'ou	Changchiak'ou N.	—	—	10	10	10	10
	Hsichihmen	Fengt'ai	—	—	13	13	13	13
	Hsichihmen	Tungpienmen	—	1	12	12	12	12
	Tungpienmen	T'iehtaoyenchiu yuan	—	—	9	9	9	9
	Hsichihmen	Panchiao	—	26	56	56	56	56
	Hsuanhua	P'angchiapu	—	—	45	45	45	45
Feng-Sha	Fengt'ai	Shach'eng	—	—	104	104	104	104
Ch'ing-Shan	Peking	Shanhaikuan	418	418	417	417	417	417
	Wuch'ing[d]	Chihsien	—	—	—	—	96	96
	Peitaiho	Haipin	—	—	10	10	10	10
	Tangku	Hsinkang	—	—	—	10	10	10
	Tangku	Tangku South	—	—	—	5	5	5
	Hanku[s]	Nanpao	—	—	—	—	—	39
Ch'ing-Ch'eng	Tungpienmen[h]	Yingsh'ouying	90	128	176	176	176	176
	Shangpanch'eng	Yingsh'ouying	—	16	61	61	61	61
	Hsuchiachan	Miaot'aitzu	—	—	11	11	11	11
Chin-Ch'eng N	Sanshihchia[b]	Ch'engte	—	45	142	142	142	142
	T'unghsien W.	T'unghsien E.	—	—	6	6	6	6
Ch'ing-P'u	Tientsin	Sangyuan	217	217	217	217	217	217
	Liangwangchuang	Huitui	—	—	26	26	26	26
Te-Shih	Palichuang	Shihchiachuang	172	172	172	172	172	172
Sheng-T'ai	Shihchiachuang	Niangtzukuan	73	73	73	73	73	73
		Total	1689	1814	2405	2449	2699	2984

APPENDIX B--Continued

Line	Terminals		Length (km) in Operation at End of Indicated Year					
	From	To	1949[a]	1952[a]	1957[a]	1963[a]	1970	1975
	PROVINCE							
	HUNAN							
Che-Kan	P'inghsiang	Chuchou	—	82	82	82	82	82
Yueh-Han	Yangloussu	P'ingshih	315	602	602	602	602	602
Hsiang-Kuei	Hengyang	Tsuch'i	187	187	187	187	187	187
Hsiang-Ch'ien	Chuchou[h]	Louti	—	—	142	142	142	142
	Louti	Chinchushan	—	—	—	65	65	65
	Chinchushan[k]	Yup'ing	—	—	—	—	—	356
Lou-Shao	Louti	Shaoyang	—	—	—	98	98	98
Ch'en-San	Ch'enhsien[h]	Santu	—	—	36	36	36	36
Wu-Pai	Wuch'i	Paishihtu	—	—	14	14	14	14
Shao-Shan	Shangshao[l]	Shaoshan	—	—	—	—	21	21
		Total	502	871	1063	1226	1247	1603
	HUPEI							
Ch'ing-Han and Yueh-Han	Wushengkuan	Yangloussu	159	347	347	347	347	347
	Hanshui	Line	—	—	6	6	6	6
	Yangtze	Line	—	—	9	9	9	9
	Huayuan[d]	Anlu	—	—	—	—	—	35
Wu-Sui	Wuch'ang	Suihsien	—	—	—	216	216	216
	Suihsien[m]	Chunhsien	—	—	—	—	204	204
	Chunhsien[n]	Paiho	—	—	—	—	—	160

Wu-Huang	Wuch'ang[h]	Huangshih	—	—	—	124	124	124
	Tieh-shan[h]	Linghsiang	—	—	—	30	30	30
	Hsiangfan[d]	Ch'ich'eng	—	—	—	—	253	253
		Total	159	347	362	732	1189	1384
	KANSU							
Lung-Hai	T'oshih	Lanchou	74	74	422	422	422	422
Pao-Lan	Kant'ang	Lanchou	—	—	—	232	232	232
Lan-Hsin	Lanchou	Weiya	—	332	942	1185	1185	1185
	Ch'ingshui[d,*]	Ojne Qi	—	—	—	—	200	200
Ti-Pai	Tichiutai	Paiyin	—	—	23	23	23	23
Lan-Ch'ing	Hok'ou	Shaich'ewan	—	—	—	53	53	53
Hei-Lao	Heishanhu[h]	Laochunmiao	—	—	21	21	21	21
Lan-A	Lanchou	Akanchen	—	—	21	21	21	21
Wu-Kan	Wuwei	Kant'ang	—	—	—	185	185	185
	Chiayukuan[d]	Paihuwantzu	—	—	—	—	66	66
		Total	74	406	1429	2142	2408	2408
	KUANGTUNG							
Yueh-Han	P'ingshih	Canton	—	334	334	334	334	334
Kuang-Chiu	Canton	Shench'uan	146	146	146	146	146	146
	Chishan	Huangpu	—	—	4	4	4	4
Kuang-San	Canton	Sanshui	—	49	49	49	49	49
Lien-Mao	Lienchiang (Hots'un)	Maoming	—	—	—	61	61	61
Li-Chan	Wenli	Chanchiang	—	—	106	106	106	106
Hai-Nan	Paso	Shihlu[c]	—	—	52	52	52	52
	Huangliu	Yulin	—	—	—	110	110	110
		Total	146	529	691	862	862	862

APPENDIX B--Continued

Line	Terminals		Length (km) in Operation at End of Indicated Year					
	From	To	1949[a]	1952[a]	1957[a]	1963[a]	1970	1975
	PROVINCE KUEICHOU							
Ch'ien-Kuei	Mawei	Kueiyang	—	157	293	293	293	293
	Kueiyang[o]	Kanshui	—	—	—	—	328	328
	Kueiyang[p]	Shuangho	—	—	—	—	309	309
Hsiang-Ch'ien	Yup'ing[k]	Kueiting	—	—	—	—	—	276
		Total	—	157	293	293	930	1206
	LIAONING							
Chin-Ch'eng	Yehpaish'ou[b]	Sanshihchia	—	73	73	73	73	73
Chin-Ch'eng	Yehpaish'ou[b]	Chihfeng	—	—	38	38	38	38
Shen-Shan	Shenyang	Shanhaikuan	420	426	426	426	426	426
	Shenyang N.	Shenyang E.	—	—	11	11	11	11
	Shenyang N	Huangkuts'un	—	3	3	3	3	3
	Yuhung	Tuch'eng	—	5	5	5	5	5
	Chinhsi	Hulutao	—	12	13	13	13	13
	Goubangzi	Panshan	—	—	—	—	32	32
Chin-Ch'eng	Chinchou	Yehpaish'ou	50	222	222	222	222	222
	Chinlingssu	Peip'iao	—	18	18	18	18	18
I-Kao	Ihsien	Kaot'aishan	132	192	192	192	192	192
Ta-Cheng	Tahushan[b]	Aerhhsiang	164	164	164	164	164	164
Ta-Shen	Luta	Shenyang	397	397	397	397	397	397

	Tienchia[h]	Wutao	—	—	—	61	61	61
	Choushuitzu	Lushun	—	50	50	52	52	52
	Choushuitzu	Kanchingtzu	—	14	14	14	14	14
	Chinchou	Ch'engtzut'ung	102	102	102	102	102	102
	Tashihch'iao	Yingk'ou	23	23	23	23	23	23
	Haich'eng[e]	Panshan	—	—	—	—	—	66
Shen-Pin	Shenyang	Ch'angt'u	136	136	136	136	136	136
	Ch'angtu[b]	Ssup'ing	53	53	53	53	53	53
	Ssup'ing[b]	Sanchiangk'ou	—	64	64	64	64	64
Shen-An	Shenyang	Antung	261	258	277	277	277	277
	Suchiats'un	Yushut'ai	—	13	13	13	13	13
	Hunho	Fushun	53	49	57	57	57	57
	Pench'i	Liaoyang	70	71	69	69	69	69
	South Pench'i	Tienshihfu	—	85	86	86	86	86
	Fenghuangch'eng	Changtien	—	112	143	143	143	143
	Kuanshui[c]	Saimachi	—	—	—	—	33	33
	Antung	Antung E.	—	13	13	13	13	13
Shen-Chi	Shenyang[b]	Chaoshih	161	161	161	161	161	161
	Fushun S.	Fushun N.	—	4	4	4	4	4
T'ieh-San	T'iehling[b]	Sanchiatzu	—	—	—	33	33	33
T'ieh-Fa	Sanchiatzu[b]	Fak'u	—	—	—	20	20	20
Nu-nan	Nuerhho	Nanp'iao	—	—	—	29	29	29
		Total	2022	2720	2827	2972	3037	3103
	SHANHSI							
Ch'ing-Pao	Hsiwanp'u	Paotzuwan	151	151	151	151	151	151
Cheng-T'ai	Niangtzukuan	Yutz'u	131	131	131	131	131	131
	Hsinching	Weishui	—	—	11	11	11	11

APPENDIX B--Continued

Line	Terminals		Length (km) in Operation at End of Indicated Year					
	From	To	1949[a]	1952[a]	1957[a]	1963[a]	1970	1975
	PROVINCE							
	SHANHSI							
	Liyuan[c]	Kutui	—	—	—	—	41	41
	Poai[c]	Wuyang	—	—	—	—	162	162
T'ung-P'u	Tat'ung	Fenglingtu	164	835	860	860	860	860
	Yuanp'ing[d]	Pingshingkuan	—	—	—	—	154	154
	Pingshingkuan	Laiyuan	—	—	—	—	—	53
	Chiangts'un	Hsihsien	—	—	31	31	31	31
	Shanglants'un	Chifufen	—	—	16	16	16	16
	P'ingwang	K'ouch'uan	—	—	10	10	10	10
	Chiehhsiu	Yangch'uanchu	—	—	—	46	46	46
	T'aiyuan	T'aohsing	—	—	—	27	27	27
	K'ouchuan	Wangts'un	—	—	—	60	60	60
	T'ungkuan	Fenglingtu	—	—	—	9	9	9
	Ningwu[d]	Wutsai	—	—	—	—	—	88
		Total	446	1117	1210	1352	1709	1850
	SHANTUNG							
Ching-Pu	Sangyuan	Hanchuang	414	414	414	414	414	414
	Yenchou[h]	Chining	—	—	—	32	32	32
	Taian[c]	Muchuang	—	—	—	—	44	44
Te-Shih	Techou	Palichuang	9	9	8	8	8	8

	Hsint'ai	Tz'uyao	—	66	67	67	67	67
	Hsint'ai[c]	Laiwu	—	—	—	—	111	111
	Hsuehch'eng	Tsaochuang	—	—	32	32	32	32
Chiao-Chi	Chinan	Ch'ingtao	393	392	393	393	393	393
	Patou	Changtien	—	—	49	49	49	49
	Changtien[d]	Choitsai	—	—	—	—	—	66
Lan-Yen	Lants'un	Yentai	—	—	183	183	183	183
		Total	816	881	1146	1178	1333	1399
	SHENHSI							
Lung-Hai	T'ungkuan	T'oshih	386	386	386	386	386	386
Pao-Ch'eng	Paochi	Tat'an	—	—	—	302	302	302
	Luehyang[d]	Ch'engku	—	—	—	—	132	132
	Ch'engku[n]	Ank'ang	—	—	—	—	—	196
T'ung-Hsien	T'ungch'uan	Hsienyang	135	135	135	135	135	135
	(**)[d]	Hancheng	—	—	—	—	—	192
Hsi-Yu	Hsian	Yuhsia	—	—	—	45	45	45
	Ank'ang[n]	Paiho	—	—	—	—	—	120
	Ank'ang[n]	Wanyuan	—	—	—	—	—	104
		Total	521	521	521	868	1000	1612
	SSUCH'UAN							
Ch'eng-K'un[b]	Ch'engtu[r]	Yungjen	—	—	—	—	749	749
Pao-Ch'eng	Tat'an	Ch'engtu	—	—	314[t]	367	367	367
	Kuangyuan[d]	Paishui	—	—	—	—	33	33
	Teyang[c]	Mienchu	—	—	—	—	34	34
Ch'eng-Yu	Ch'engtu	Chungking	—	504	504	504	504	504

APPENDIX B--Continued

Line	Terminals		Length (km) in Operation at End of Indicated Year					
	From	To	1949[a]	1952[a]	1957[a]	1963[a]	1970	1975
	PROVINCE SSUCH'UAN							
	Chungking[d]	Senhuipa	—	—	—	—	75	75
	Sanhuipa	Wanyuan	—	—	—	—	—	310
Ch'uan-Ch'ien	Hsiaonanhai	Kanshui	58	58	119	119	119	119
San-Wan	Sanchiang	Wansheng	—	—	32	32	32	32
Nei-K'un	Neichiang	Anpien	—	—	—	142	142	142
Hsin-Kuan	Hsintu[h] (ch'ingpaichiang)	Kuanhsien	—	—	—	51	51	51
		Total	58	562	969	1215	2106	2416
	YUNNAN							
Nei-K'un	K'unming	Chani	174	174	180	180	180	180
	Chani[p]	Shuangho	—	—	—	—	154	154
	Yentang[e]	Tachungshu	—	—	—	—	—	35
	Chani[e]	Fuyuan	—	—	—	—	—	70
Tien-Yueh	K'unming	Hok'ou	—	—	469	469	469	469
Pi-Shih	Pisechai	Shihp'ing	—	—	143	143	143	143
Ko-Chi	Kochiu	Chichien	—	—	34	34	34	34
Ch'eng-K'un[b]	K'unming	Ip'inglang	—	—	—	137	137	137
	Ip'inglang[d,r]	Yungjen	—	—	—	—	199	199
K'un-Shih	K'unming	Shihtsui	12	12	12	12	12	12
		Total	186	186	838	975	1328	1433

	AUTONOMOUS REGIONS							
	HSINCHIANG							
Lan-Hsin	Weiya	Urumchi	—	—	—	707	707	707
	Urumchi[d]	Fuk'ang	—	—	—	—	88	88
		Total	—	—	—	707	795	795
	INNER MONGOLIA							
	Shuangch'engtzu[d,*]	Ulannur	—	—	—	—	84	84
	Shuangch'engtzu[d,*]	Ojne Qi	—	—	—	—	170	170
Ta-Cheng	Aerhhsiang*	Chengchiats'un	202	202	206	206	206	206
	Kouk'ou*	125 km	—	—	126	126	126	126
Ching-Pao	Paotzuwan	Paot'ou	211	420	420	420	420	420
Chi-Erh	Chining	Erhlien	—	—	330	330	330	330
	Gurban Obo	Quagan Nur	—	—	—	—	—	60
Ch'ang-A	Kokenmiao*	Aerhshan	283	283	283	283	283	283
Pao-Lan N.	Paot'ou	Sanshengkung	—	—	—	278	278	278
	Sanshengkung[b]	Wuta	—	—	—	110	110	110
	Wuta*	Jarantai	—	—	—	—	121	121
Pao-Pai	Paot'ou[h]	Paiyunopo	—	—	158	158	158	158
Erh-Shui	Erhtaoshaho[h]	Shuimot'an	—	—	30	30	30	30
Chin-Ch'eng N.	T'ieni[b,*]	Ch'ihfeng	—	—	109	109	109	109
	Haipowan[e]	Raseng Miao	—	—	—	—	—	46
Sui-Man N.	Nientzushan*	Manchouli	579	579	580	580	580	580
Ya-Lin	Yak'oshih*	K'utuerh	—	—	144	144	144	144
	K'utuerh[h,*]	Kanho	—	72	221	221	221	221
	Kanho[d,*]	Jagdaqi	—	—	—	—	88	88
	Chaochung[d,e,*]	Mordaga	—	—	—	32	87	87

APPENDIX B--Continued

Line	Terminals		Length (km) in Operation at End of Indicated Year					
	From	To	1949[a]	1952[a]	1957[a]	1963[a]	1970	1975
	AUTONOMOUS REGIONS							
	INNER MONGOLIA							
	It'uliho*	Kenho	–	–	27	27	27	27
	Kenho[d,e,*]	Mankuei	–	–	–	81	169	169
	T'aip'ingch'uan[d,*]	Tungliao	–	–	–	–	110	110
	Nenchiang[d,*]	Jagdaqi	–	–	–	–	160	160
	Jagdaqi[e,*]	Linhai	–	–	–	–	188	188
		Total	1275	1556	2634	3135	4199	4305
	KUANGHSI							
Hsiang-Kuei	Tzuch'i	Munankuan	346	838	826	826	826	826
	Liuchou[e,s]	Lat'ung	–	–	–	–	–	36
Li-Chan	Lit'ang	Wenli	–	–	209	209	209	209
Ch'ien-Kuei	Liuchou	Mawei	161	161	314	314	314	314
	(**)[e]	Locheng	–	–	–	–	–	40
		Total	507	999	1349	1349	1349	1425
	NINGHSIA							
Pao-Lan	Wuta[b]	Kant'ang	–	–	–	360	360	360
	Pinglo[e]	Shitanjing	–	–	–	–	–	36
	Wuta[d,*]	Jarantai	–	–	–	–	–	–
	Yingchuan Station[d]	Yinchuan	–	–	–	–	18	18
		Total	–	–	–	360	378	414
		Grand Total	17137	22554	28993	34091	40725	43982

Notes:

a Figures mainly taken from Wu, Yuan-li, The Spatial Economy of Communist China (New York: Frederick A. Praeger, 1967), table E-1, pp. 252-261. Figures rearranged to coincide more accurately with provincial areas, or modified where necessary are indicated.

b Re-arranged of distribution of Wu's data with reference to (1) Communist China Railroad Passenger Timetable July - Nov. 1963, JPRS #21963, U.S. Department of Commerce, and/or (2) Atlas, People's Republic of China, C.I.A., U.S. Government, Nov. 1971.

c Communist China Administrative Atlas, U.S. Government, March 1969, Figures obtained by measurement.

d Atlas, People's Republic of China, C.I.A., U.S. Government, Nov. 1971. Figures obtained by measurement using other large-scale maps and atlases available.

e Zhonghua Remin Gongheguo Ditu [Map of the People's Republic of China], 1:6,000,000, Peking: June 1974, first edition, first printing. Figures obtained by measurement using other large-scale maps and atlases available.

f This line, when fully rebuilt by 1957 terminates at Ningpo instead of Ch'uan-shan as shown in Wu, Yuan-li, op. cit., 1967.

g People's Daily, Peking, February 10, 1972.

h Figure(s) taken from, or modified in accordance with, Communist China Railroad Passenger Timetable, 1963, op. cit.

i Chung-hua Jen-min kung-ho-kuo ti-tu [Map of the People's Republic of China], 1:6,000,000, Peking: 1965 edition, 7th printing, Jan. 1973.

k Ta Kung Pao, Hong Kong, Sept. 26, 1974 and Mar. 17, 1975.

l People's Daily, Peking, February 11, 1967.

m Yang-ch'eng Wen Pao [Canton Evening Post], Canton, January 2, 1966 and People's Daily, Peking, January 9, 1966.

n Joint Economic Committee, U.S. Congress, China: A Reassessment of the Economy, July 1975, p. 267.

o Wen Hui Pao, Shanghai, February 24, 1957 and Ta Kung Pao, Hong Kong, March 17, 1975.

Notes:

p Ta Kung Pao, Hong Kong, December 23, 1974.

q Ta Kung Pao, Hong Kong, July 6, 1969.

r Ta Kung Pao, Hong Kong, March 23, 25, and 31, 1974 and A Glance at China's Economy, Peking, 1974, p. 32.

s Chung-hua jen-min kung-ho-kuo fun sh'eng ti-tu [Provincial Atlas of the People's Republic of China] Peking, Ti-tu Chu-p'an she, October 1974.

t Chuan-Kou lu-ka lieh-che shi-ka piao [National passenger train timetable], Peking: People's Railways Press. 11, May, 1956, shows that passenger services were already available between Ch'eng-tu and Kuang-yuan, though the line was probably completed as far as Yang-ping.

* Re-arrangement of data in accordance with provincial boundary change in 1969.

** Exact location and place name not yet verified.

APPENDIX C

RAILWAY ACCESSIBILITY RANKINGS, 1949-1975

Selected Cities	1975		1970		1963		1957		1949	
	SPA[a]	Rank	SPA[a]	Rank	SPA[a]	Rank	SPA[a]	Rank	SPA[a]	Rank
Shihchiachuang	2466	1	1850	1	1513	4	1402	9	481	13
Techou	2531	2	1909	3	1507	2	997	4	436	8
Peking	2535	3	1862	2	1512	3	1015	6	454	9
Tientsin	2566	4	1921	4	1506	1	970	2	409	5
T'ungkuan	2578	5	1978	6	1668	10	—	—	—	—
Chengchou	2678	6	2049	9	1682	11	1211	21	555	22
Hsinhsiang	2682	7	2181	8	—	—	—	—	—	—
Yehpaish'ou	2683	8	1955	5	1578	5	1023	8	—	—
Chinan	2685	9	2030	7	1580	6	1048	10	473	12
Loyang	2687	10	2110	13	—	—	—	—	—	—
Hsian	2740	11	—	—	1803	18	1336	31	622	32
T'aiyuan	2757	12	2087	12	1735	12	1161	16	—	—
Hsuchou	2829	13	2181	17	1653	8	1166	17	510	16
Mengmiao	2836	14	2171	16	1803	17	1308	29	—	—
Chinchou	2836	15	2059	10	1605	7	956	1	394	3
Tat'ung	2881	16	2164	14	1789	15	1281	26	521	18
Hsinlits'un	2902	17	2079	11	1664	9	976	3	422	6
Chihfeng	2947	18	2169	15	1761	14	1156	15	—	—
Hsiangyang	3012	19	—	—	—	—	—	—	—	—
Paochi	3032	20	2332	22	2072	34	—	—	—	—
Huhehot	3068	21	2317	21	1924	23	1410	38	592	30

APPENDIX C--Continued

Selected Cities	1975		1970		1963		1957		1949	
	SPA[a]	Rank	SPA[a]	Rank	SPA[a]	Rank	SPA[a]	Rank	SPA[a]	Rank
Lienyunkang	3093	22	2395	25	1836	20	1299	28	581	27
Wuhan	3123	23	2287	19	1924	24	1402	36	626	34
Haich'eng	3183	24	—	—	—	—	—	—	—	—
Yanpingkuan	3220	25	—	—	—	—	—	—	—	—
Chengchiats'un	3223	26	2191	18	1755	13	1004	5	—	—
Paot'ou	3229	27	2445	27	2059	31	—	—	—	—
Shenyang	3249	28	2348	24	1833	19	1017	7	376	1
Pangfou	3263	29	2445	28	1802	16	1279	25	565	25
Ch'angsha	3268	30	2405	26	2050	29	1405	37	—	—
Erhlien	3332	31	2531	31	2107	36	1543	49	—	—
Yinch'uan	3371	32	2556	33	2148	39	—	—	—	—
Ssup'ing	3374	33	2307	20	1852	21	1058	11	397	4
Paich'engtzu	3415	34	2336	23	1870	22	1069	12	—	—
Ch'ingtao	3463	35	2660	42	2117	37	1435	41	544	20
Yent'ai	3463	36	2660	43	2117	38	1435	42	—	—
Nanking	3470	37	2619	36	1941	25	1384	34	624	33
Kant'ang	3476	38	2656	39	—	—	—	—	—	—
Lanchou	3477	39	2705	47	2390	56	1465	44	—	—
Paiyunopo	3493	40	2659	40	2242	46	1543	50	—	—
Hofei	3527	41	2659	41	1985	27	1412	39	636	35
Ch'engtu	3528	42	2566	34	2239	45	—	—	—	—
Ch'angch'un	3547	43	2459	29	1977	26	1133	13	426	7
Angangch'i	3606	44	2479	30	1986	28	1135	14	—	—

Louti	3620	45	–	–	2229	44	–	–	–	–
Meihok'ou	3639	46	2652	38	2064	32	1189	18	465	11
Shanghai	3649	47	2753	49	2068	33	1481	45	687	37
Hengyang	3937	48	2621	37	2342	53	1619	55	–	–
Aerhshan	3679	49	2550	32	2053	30	1202	20	–	–
Kunming	3684	50	2678	45	–	–	–	–	–	–
Wuwei	3714	51	2856	56	2321	49	–	–	–	–
Kueiting	3756	52	–	–	–	–	–	–	–	–
Harbin	3757	53	2610	35	2088	35	1192	19	459	10
Hangchou	3807	54	2875	57	2192	42	1578	52	691	38
Fuyu	3838	55	2674	44	2153	40	1252	22	–	–
Kouk'ou	3856	56	2683	46	2157	41	1256	23	–	–
Kueiyang	3910	57	2763	50	2696	75	1877	66	–	–
Canton	3937	58	2831	55	2521	67	1748	62	–	–
Pisechai	3940	59	2884	59	–	–	–	–	–	–
Hsining	3945	60	3089	70	2547	68	–	–	–	–
Ch'ingshui	3966	61	–	–	–	–	–	–	–	–
Lafa	3991	62	2791	52	2218	43	1276	24	–	–
Liuchou	3996	63	2945	48	2513	64	1744	60	–	–
Mutanchiang	3996	64	2799	53	2397	58	–	–	506	14
Peian	4023	65	–	–	–	–	–	–	591	28
Antung	4048	66	3140	75	2513	65	1503	46	565	24
Chiakutach'i	4084	67	–	–	–	–	–	–	–	–
Chinhua	4088	68	2987	64	2313	48	–	–	–	–
Yak'oshih	4102	69	2884	60	2332	51	1381	32	–	–
Nanch'ang	4106	70	3009	66	2328	50	1615	54	754	41
Chungking	4107	71	3104	74	2768	79	–	–	–	–

APPENDIX C--Continued

Selected Cities	1975		1970		1963		1957		1949	
	SPA[a]	Rank	SPA[a]	Rank	SPA[a]	Rank	SPA[a]	Rank	SPA[a]	Rank
Chian	4159	72	3072	69	2426	61	1451	43	603	31
Yingt'an	4169	73	3067	67	2392	57	1649	58	—	—
Hok'ou	4204	74	3098	72	—	—	—	—	—	—
T'umen	4206	75	2958	62	2375	55	1383	33	556	23
Ningpo	4214	76	3089	71	2375	54	1711	59	—	—
Shuangch'engtzu	4226	77	—	—	—	—	—	—	—	—
Nanch'a	4240	78	3006	65	2422	60	1426	40	577	26
Lit'ang	4252	79	2951	61	2688	72	1873	65	—	—
It'uliho	4334	80	2876	58	2511	63	1510	47	—	—
Manchouli	4366	81	3098	73	2515	66	1514	48	530	19
Waiyang	4425	82	3273	80	2567	69	—	—	—	—
Lushun	4463	83	3747	86	3021	83	1746	61	—	—
Chiamussu	4488	84	3200	77	2731	76	1636	56	—	—
Ochinach'i	4490	85	—	—	—	—	—	—	—	—
Urum'chi	4490	86	3070	68	2504	62	—	—	—	—
Wuiling	4504	87	—	—	—	—	—	—	—	—
Nanning	4516	88	3165	76	—	—	2006	68	—	—
Suifenho	4518	89	3221	79	2757	78	—	—	640	36
Kulien	4608	90	—	—	—	—	—	—	—	—
Fuchou	4689	91	3487	84	2750	77	—	—	—	—
Tungfanghung	4768	92	3421	83	—	—	—	—	—	—
Chanchiang	4776	93	3375	82	3050	84	2006	67	—	—
Hsiamen	4949	94	3697	35	2929	82	1782	64	—	—

Neichiang	—	—	2770	51	2410	59	—	—	—	—
Suihua	—	—	2805	54	2255	47	1309	30	520	17
Anpien	—	—	2984	63	2593	70	—	—	—	—
Tungt'ang	—	—	3220	78	—	—	—	—	—	—
Mankuei	—	—	3300	81	—	—	—	—	—	—
Nenchiang	—	—	—	—	2336	52	1385	35	—	—
T'angwangho	—	—	—	—	2605	71	—	—	—	—
Chinho	—	—	—	—	2694	73	—	—	—	—
Kanho	—	—	—	—	2694	74	1643	57	—	—
P'inghsiang	—	—	—	—	2871	80	—	—	825	42
Hut'ou	—	—	—	—	2924	81	—	—	—	—
Fenglingtu	—	—	—	—	—	—	1294	27	—	—
Hsinch'ing	—	—	—	—	—	—	1559	51	—	—
Chiaowan	—	—	—	—	—	—	1598	53	—	—
Mishan	—	—	—	—	—	—	1779	63	697	40
Tahushan	—	—	—	—	—	—	—	—	383	2
Hsiaokuchia	—	—	—	—	—	—	—	—	509	15
Yutzu	—	—	—	—	—	—	—	—	552	21
Ningwu	—	—	—	—	—	—	—	—	592	29
T'ienshui	—	—	—	—	—	—	—	—	693	39

[a]SPA: shortest path accessibility

For method of calculation, see Alfonso Shimbel, "Structural Parameters of Communication Networks," Bulletin of Mathematical Biophysics, 15 (1953), pp. 501-507, or E. J. Taaffe and H. L. Gauthier, Jr., Geography of Transportation (Englewood Cliffs: Prentice-Hall, 1973), pp. 116-158.

APPENDIX D

NOTES AND SOURCES FOR ILLUSTRATIONS

Figure

1. Wu To, "Chin-t'ung t'ieh-lu ti cheng-i," [The debate over the Tientsin-Tungchou Railway], reprinted in Chung-kuo chin-t'ai shih lun-chung, Series I, Vol. 5, pp. 135-170.

2. Based on Ling Hung-hsun, Chung-kuo t'ieh-lu chih, op. cit., 1954. "Railway sphere of influence" is measured approximately 100 km. from either side of a railway track.

3. Chou Shih-t'ang, Erh-shih shih-chi chung wai ta ti-t'u [Map of 20th century China and the world], (n.p., 1966).

4. [a]C. T. Chow, "China's Internal Transport Problem: The Case of the Railway's First Century, 1866-1966" (Ph.D. Dissertation, Michigan State University, Department of Geography, 1972), map 11 and table 19.

   [b]Wang K'ai-chieh, Chung-kuo chin pe nien chiao-tung shih [History of communications of the last one hundred years of China] (Taipei: Chung-Kuo chiao-tung chien-she hsueh-hui), 1961.

   [c]A. Herrmann, An Historical Atlas of China (Chicago: Aldine, 1966), p. 53, partly revised edition, N. Ginsburg, ed.

   [d]T'ieh-lu hsieh-hui, ed., Min-kuo t'ieh-lu i-nien shih [Railway during the first year of the Republic], Peking: Chinghua, 1914.

5. [a]Yu Fei-p'eng, Shih-wu-nien lai chi chiao-tung k'ai-fang [Communications in the last fifteen years] (n.p., 1946).

   [b]Chung-hua tsui hsin ti-tu [Newest map of China] (n.p., n.d.).

   [c]A. Herrmann, op. cit., p. 54.

   [d]For provincial capitals and other major cities, see A. Herrmann, op. cit., and Min Pe-ch'ien, Hsin-pien chung-kuo chuan-tu [New wall map of China] (n.p., 1939).

   [e]Norton S. Ginsburg, "Manchurian Railway Development", Far Eastern Quarterly, August 1949, pp. 398-411, Fig. 2.

   [f]Chin Shih-hsuan, ed. Chung-Kuo tung-pei t'ieh-lu wen-ti hui-lun [China's northeastern railway problems] (Tientsin: Ta Kung Pao, 1932), p. 74, Map.

[g]Japan Manchoukuo Yearbook, 1934, p. 615, Map.

6. [a]Yu Fei-p'eng, op. cit., map 2.

[b]Wang K'ai-chieh, op. cit.

[c]Distance of some short, local lines not readily available were obtained by measurement. All such measurements were based on Chang Ch'i-yun, Chung-hua-min-kuo ti-t'u chi [Atlas of the Republic of China], Taipei: Institute of National Defense, 1959.

[d]U.S. Military Intelligence Division, U.S. Army, Railroads of China, Vols. 1-7 (Washington: Strategic Engineering Study, SES 145, Nov. 1944).

[e]Fu Chueh-chin, Chung-hua min-kuo hsing-cheng ch'u-yu t'u [Administrative map of the Republic of China] (n.p. 1947).

[f]Wang K'ai-chieh, op. cit.

8. Hsiao Liang-lin, China's Foreign Statistics, 1864-1949 (Cambridge, Mass.: Harvard University Press, 1974).

15. PRC: Handbook of Economic Indicators, C.I.A. Research Aid, ER 76-10540, August 1976.

16,19. Ch'uan-kuo t'ieh-lu lu-k'o lieh-ch'e shih-k'o piao, [National railway passenger train timetable] (Peking: Railway Publishing House, May 11, 1956).

17,20. Ch'uan-kuo t'ieh-lu lu-k'o lieh-ch'e shih-k'o piao, July-November, 1963 [National railway passenger train timetable, July-Nov., 1963] (Peking: Railway Publishing House, 1963). See also JPRS No. 21963 for an English translation.

18. See sources for figures 16, 17, 19, and 20.

21. [a]Chung-kuo ti-shing, 1:4,000,000 (wall map) [Relief map of China] (Peking: Ti-tu chu-pan she), March, 1974.

[b]"The Distribution of Earthquake zone in China," Ti-li chi-shi, No. 3, 1975, pp. 1-4 and 19.

[c]"China's Topography," China Reconstructs, September 1971.

[d]Chien-ming chung-kuo ti-li [A simple geography of China] (Shanghai: Jen-min chu-pan she, 1974), pp. 1-3.

[e]Hostile land includes marshy areas (marshland and marine flooded areas), desert (sandy and rocky deserts), earthquake zones, and areas with abrupt change of relief or extremely high altitude.

For the last classification, China is divided into two halves. In the low-lying eastern half of China comprising Kuanghsi, Kuangtung, Fuchien, Chechiang, Chiangsu, Shantung, Hupei, Liaoning, Chilin, Heilungchiang, Honan, Anhui, Chianghsi, Hupei, Hunan, Ssuch'uan, Kueichou, and Yunnan, all areas above the 500 meter contour line are classified as areas with abrupt change of relief. In the western half of China where the general relief is already far above 500 m. contour, change of relief is not considered a significant factor. Rather the sheer altitude renders productive activities difficult. Thus in this other half of China, all areas above 2000 meters are classified as hostile land.

APPENDIX E

THE INDEPENDENT VARIABLES: SOURCES AND EXPLANATIONS

Code	Sources and Explanations
CULTA	[a]Cultivated land area in 1957 in sq. km.
	[b]D. H. Perkins, Agricultural Development in China, 1368-1968 (Chicago: Aldine, 1969).
	[c]Data in shih-mou are converted into sq. km. at the rate of 1500=1 sq. km., and rounded off to the nearest 100. Figure for Tibet is estimated.
GRAOP, GRAPP	[a]Average grain output and average grain output per capita, 1952-59, respectively.
	[b]Both derived from Appendixes D and E in F. C. Teiwes, "Provincial Politics in China: Themes and Variations," in J. Lindbeck (ed.), China Management of a Revolutionary Society (Seattle: University of Washington Press, 1971), pp. 116-189.
	[c]Figures in original tables for Kansu included Ninghsia. For the analyses in this work, same figures are used for both units.
GVIO, GVIOP	[a]Gross value of industrial output and gross value of industrial output per capita, for the years ending 1957, 1963, 1970, and 1973.
	[b]Wu Yuan-li, The Spatial Economy of Communist China, (New York: The Hoover Institution, 1967); and R. M. Field, N. R. Lardy, and J. P. Emerson, A Reconstruction of the Gross Value of Industrial Output by Province in the PRC: 1949-73 (Washington: Department of Commerce, July 1975).
	[c]For the 1963 analysis, data from Wu have been applied.
	[d]For the analysis of other years, the relatively incomplete data from Field, et al., have been expanded, through interpolation and other methods, into complete sets which have been cross-checked against national data and against those from Wu.
OPRO	[a]Operating roads in kilometers.

OPRO

[b]Jen-min shou-ts'e, 1950, pp. ch'ih 34-36; and China Transportation Map, Foreign Economic Administration, Far East Economy Div., October 9, 1944.

[c]Where a road runs across provincial boundaries, the total length are apportioned arbitrarily after consulting various atlases and maps.

[d]For Heilungchiang, Chilin, Liaoning, and Inner Mongolia, a total of 8,520 km. of operating roads was available, from which estimated kilometrages are assigned to the four areas in accordance with observed distribution in early 1950's.

POCUL

[a]Population per unit of cultivated area of 1957, 1963, 1970, and 1974.

[b]Obtained by the formula: $\frac{POP}{CULTA}$.

PODEN

[a]Population density of 1957, 1963, 1970, and 1974.

[b]Obtained by the formula: $\frac{POP}{AREA}$.

POP

[a]Population of 1957, 1963, 1970, and 1974.

[b]J. S. Aird, Population Estimates for the Provinces of the PRC, 1953-1974 (Washington: Department of Commerce, February 1974).

[c]1957 population figures from Wei ta ti shih-nien [Great ten years], p. 9. Figures for Hopei, Kansu and Ninghsia are 1958 figures in accordance with administrative division.

[d]1963 and 1970 population data are mid-year figures, while those of 1974 are January figures, the latest obtainable from Aird. Where necessary, data have been adjusted to conform with provincial areas of the mid-1960's.

TRANR

[a]Combined rankings by OPRO and WAWAY, operating roads and navigable waterways, both in kilometers.

URPOP,
URDEN

[a]Urban population, and density of urban population 1953 and 1958.

[b]E. Ni, Distribution of the Urban and Rural Population of Mainland China, 1953 and 1958 (Washington: Bureau of Census, October, 1960).

[c]Urban population figure for Kansu (including Ninghsia) has been adjusted to obtain separate estimates for the two units.

[d]Density of urban population is obtained by the formula: $\frac{URPOP}{AREA}$.

WAWAY

[a]Navigable waterways in kilometers.

[b]Jen-min shou-ts'e, 1950, pp. ch'ih 32-34; and Wang Kuang, Chung-kuo shui-yun ch'ih, (Taipei: Chung-hua ta-tien pein-yin hui, 1966), various tables.

[c]Lengths of major waterways are generally taken from _Jen-min shou-ts'e_, except those of the Yangtzu, Han Shui, Hsi Chiang, Chia-ling Chiang, which run across provincial boundaries. In these cases, lengths of waterways are taken from tables in Wang Kuang in accordance with current provincial areas concerned. The resultant discrepancy amounts to 4.69% of the total waterway kilometrage.

AREA

[a]Total land area in square kilometers.

[b]T. Shabad, _China's Changing Map_ (New York: Praeger, 1972).

[c]Data in square miles are converted into square kilometers by the formula 1 sq. mile = 2.589988 km^2 and rounded off to the nearest 100.

DEGRE, DEGRN

[a]Degrees of longitude east, and degrees of latitude north.

[b]Readings are taken of provincial or autonomous regional capitals from _Chung-kuo ti-t'u ts'e_ ("Atlas of China") (Peking: Ti-t'u chu-pan she, 1974).

HOSTI

[a]Per cent of hostile land area.

[b]For sources and explanations, see appendix D and Figure 21.

SHAPE

[a]Shape of province or autonomous region.

[b]Following Kansky, op. cit. (1963), p. 45. Shape is obtained by the formula: $S = \frac{A_1}{A_p}$, where S is shape, A_1 is the air line distance along the longest axis, and A_p is the air line distance along the perpendicular. Measurements are taken from _China Administrative Atlas_, (Washington: C.I.A., March 1969), and _Chung-hua jen-min kung-ho-kuo ti-t'u, 1:6,000,000_ (Peking: Ti-tu ch'u-pan she, 1973).

ARDIS

[a]Air distance from Peking and Shanghai, the political core area.

[b]Measurements taken from provincial or autonomous regional capitals and based on _Chung-hua jen-min kung-ho-kuo ti-t'u, 1:6,000,000_, op. cit., 1973.

CCPMB

[a]Chinese Communist Party membership, mid-1956.

[b]F. C. Teiwes, op. cit., table 9.

[c]Figures for Kansu and Ninghsia adjusted.

LOCIN

[a]Index of "geopolitical" location.

[b]The index is arrived at through a fairly complicated mathematical process. Briefly, it takes into account: (1) the location of the province or autonomous region; (2) the length of the international boundary; (3) whether it is adjacent to a land neighbor or ocean; (4) whether the adjacent neighbor is a superior or inferior power; (5) the type of disputes, if any, that have occurred along the international boundary and beyond.

RSTAB [a]Stability rank of provincial party secretaries, 1956-66.

[b]F. C. Teiwes, op. cit., table 3

PHAPC [a]Per cent of farm households in Higher Agricultural Producers' Cooperatives, mid-1956.

[b]F. C. Teiwes, op. cit., table 11.

[c]Same percentage for Kansu is adopted for Ninghsia.

PMIN [a]Per cent of non-Han population, 1953.

[b]Based on L. A. Orleans, Every Fifth Child (London: Eyre Methuen, 1972), table 10a and map, pp. 110-111. All figures, where not available directly from table, are taken at mid-point of percentages shown in map. Figures for Kansu and Ninghsia have been adjusted.

GLOSSARY

This Glossary follows a slightly modified Wade-Giles system; that is, to avoid confusion with numerous railway lines mentioned in the text, all place names are not hyphenated. A few extremely well known places or personal names, e.g., Peking, Harbin, Chiang Kai-shek, etc., are used in their conventional forms.

Personal names include only those mentioned in the text but not necessarily others footnoted.

Wade-Giles	Chinese
Aerhshan	阿爾山
Aihui	瑷琿
Amoy (see Hsiamen)	厦門
Angangch'i	昂昂溪
Anhui	安徽(省)
Ank'ang	安康
Antung	安東(丹東)
Arthur, Port (see Lushun)	旅順
Canton (Kuangchou)	廣州
Chanchiang	湛江
Chang Chih-tung	張之洞
Ch'angchiak'ou	張家口
Changchou	漳州
Ch'angch'un	長春
Changling	樟嶺
Changp'ing	漳平
Ch'angsha	長沙
Chani	沾益
Ch'aochou	潮州
Ch'aoyang	潮陽
Che (see Chechiang)	浙
Chechiang	浙江(省)
Chefoo (see Yent'ai)	煙台
Ch'eng Teh-chuan	程德全
Ch'engchiats'un	雙遼(鄭家屯)
Chengchou	鄭州
Chengte	承德
Ch'engtu	成都
Chennankuan	鎮南關
Chiahsing	嘉興
Chiamussu	佳木斯
Chian	集安
Chiang Kai-shek	蔣介石
Chianghsi	江西(省)

Chiangsu	江蘇(省)
Chiangmen	江門
Chiaochou	膠州
Chiaotso	焦作
Ch'ichiang	綦江
Ch'ich'ihaerh	齊齊哈尔
Ch'ien (see Kueichou)	黔
Chienhokou	澗河口
Chifufen	七府坟
Ch'ihfeng	赤峰
Chihli	直隸
Chilin	吉林(省,市)
Chinan	濟南
Chinch'engchiang	金城江(河池)
Chinchou	錦州
Ch'ingchiang	清江
Ch'inghai	青海(省)
Ch'ingshui	清水
Ch'ingtao	青島
Chining	集寧
Chiuchiang	九江
Chongjin	清津
Ch'uan (see Ssuch'uan)	川
Ch'uchiang	曲江
Chuchou	株州
Ch'un, Prince	醇親王
Chungking (Ch'ungch'ing)	重慶
Dairen (see Talien)	大連
Erhlien	二連浩特
Feng Chun-kuang	馮焌光
Fenglingtu	風陵渡
Fengt'ai	豐台
Fuchien	福建(省)
Fuchou	福州
Fushun	撫順
Fuyuan	富源
Haerhpin (see Harbin)	哈尔濱
Hailun	海倫
Hailung	海龍
Hami	哈密
Hangchou	杭州
Hankow (Hank'ou)	漢口
Hantan	邯鄲
Harbin (Haerhpin)	哈尔濱
Heiho	黑河
Heilungchiang	黑龍江(省)
Hengyang	衡陽
Hofei	合肥
Hokang	鶴崗
Honan	河南(省)
Hopei	河北(省)
Hsiamen	厦門
Hsian	西安
Hsiang (see Hunan)	湘
Hsiangfan	襄樊
Hsiangt'an	湘潭
Hsiangyang	襄陽
Hsiaonanhai	小南海
Hsienyang	咸陽
Hsihsien	歙縣
Hsinchiang	新疆(自治區)
Hsinhsiang	新鄉
Hsining	西寧
Hsu Shih-ch'ang	徐世昌

Hsuchou	徐州
Hsufu	敘府
Hsukuochuang	胥各莊
Huan (see Anhui)	皖
Huang (River)	黄河
Huining	會寧
Hulutao	葫蘆島
Hunan	湖南(省)
Hupei	湖北(省)
Hut'ou	虎頭
Inner Mongolia	内蒙古(自治區)
Ip'inglang	平浪
Ishan	宜山
Jeho	熱河(省)
K'aifeng	開封
Kaip'ing	開平
K'an (see Chianghsi)	贛
Kan Wang	干王(洪仁玕)
K'ang Yu-wei	康有為
Kanshui	趕水
Kansu	甘肅(省)
K'eshan	克山
Kochiu	個舊
Kuang Hsu, Emperor	光緒(皇帝)
Kuanghsi	廣西(自治區)
Kuangtung	廣東(省)
Kuan-shang ho-pan	官商合辦
Kuan-tu shang-pan	官督商辦
Kuei (see Kuanghsi)	桂
Kueichou	桂州
Kueilin	桂林
Kueisui	歸綏
Kueiting	貴定
Kueiyang	貴陽
K'unming	昆明
Kuyeh	古冶
Lafa	拉法
Laipin	來賓
Lanchou	蘭州
Lants'un	藍村
Li Hung-chang	李鴻章
Liang Ch'i-ch'ao	梁啟超
Liaoning	遼寧(省)
Link'ou	林口
Lit'ang	黎塘
Liu K'un-yi	劉坤一
Liu Ming-chuan	劉銘傳
Liuchou	柳州
Loyang	洛陽
Luan	潞安
Lushun	旅順
Luta	旅大
Lut'ai	蘆台
Manchoukuo	滿州國
Manchouli	滿州里
Mao Tse-tung	毛澤東
Meihok'ou	梅河口(海龍)
Mengting	勐定
Mien (Burma)	緬
Min (see Fuchien)	閩
Min ying	民營
Minhou	閩侯(福州)
Munankuan	友誼關
Mutanchiang	牡丹江

Nanch'ang	南昌
Nanking	南京
Nanning	南寧
Nanp'ing	南平
Nenchiang	嫩江
Ninghsia	寧夏(自治區)
Ningpo	寧波
Ningwu	寧武
Paich'engtzu	白城子
Paochi	寶雞
Paot'ou	包頭
Peihai	北海
Peip'iao	北票
Peip'ing (see Peking)	北平
Peking	北京
P'inghsiang	憑祥
Pisechai	碧色寨
Puchou	蒲州
P'uk'ou	浦口
Sanshui	三水
Shach'eng	沙城
Shanghai	上海
Shanhaikuan	山海關
Shanhsi	山西(省)
San-min chu-i	三民主義
Shant'ou	汕頭
Shantung	山東(省)
Sheng Hsuan-huai	盛宣懷
Shenhsi	陝西(省)
Shenyang	沈陽
Shihchiachuang	石家莊
Shihlung	石龍

Shouhsien	朔縣
Solun	索倫
Ssuch'uan	四川(省)
Ssup'ing (or Ssup'ingchieh)	四平
Suihsien	隨縣
Suiyuan	綏遠
Sun Yat-sen	孫逸仙(中山)
Swatow (see Shant'ou)	汕頭
Tahushan	大虎山
Taip'ing Heavenly Kingdom	太平天國
T'aiyuan	太原
Taku	大沽
Talien	大連
Talitzu	大栗子
Tanchiangk'ou	丹江口
T'ang T'ing-shu	唐廷樞
T'angshan	唐山
Taok'ou	道口
T'aonan	洮南
Tat'ung	大同
Tayeh	大冶
Tehsien	德縣
Tibet	西藏(自治區)
Tien (see Yunnan)	滇
T'ienchiaan	田家庵(淮南)
T'ienshui	天水
Tientsin (Tienchin)	天津
Tihua (see Urumchi)	迪化
Tolun	吐倫
Tso Tsung-t'ang	左宗棠
Tsungli Yamen	總理各國事務衙門

T'ungchou	通州
T'unghsien	通縣
T'ungkuan	潼關
T'ungliao	通遼
Tunhua	敦化
T'umen	圖們
Tuyun	都匀
Urumchi (Wulumuch'i)	烏魯木齊
Wang Chih-ch'un	王之春
Wang T'ao	王韜
Weihaiwei	威海衛
Weining	威寧
Woosung (Wusung) Railway Co.	滬淞鉄路公司
Wuhan	武漢
Wusung	吳淞
Yangch'u	陽曲
Yangp'ingkuan	陽平關
Yangtzu (River)	揚子(江)
Yehpaish'ou	葉伯壽(建平)
Yenchuang	閻莊
Yent'ai	煙台
Yi Ho Tuan	義和團
Yingt'an	鷹潭
Yuan Shih-k'ai	袁世凱
Yuanp'ing	原平
Yueh (see Kuangtung)	粵
Yuch'ik'ou	裕溪口
Yung Hung	容閎
Yungan	永安
Yunnan	雲南(省)

SELECTED BIBLIOGRAPHY

I. PRIMARY SOURCES

A. Books

Barker, John E. Chinese Railway Accounts. (n.p., probably around 1915).

Bland, J. O. P. Memorandum on Railway Construction in China. Shanghai, n.p., 1906.

Bourne, T. J. Construction of the Lu Han Railway. Institution of Civil Engineers, n.d.

Cary, Clarence. China's Present and Prospective Railways. New York: n.p., 1899.

Chang Chih-tung. Chang wen hsiang kung ch'uan chi [Complete works of Chang Chih-tung], Peiping: Man-hua-tsai, n.d.

________. Chang wen-hsiang-kung ts'ou-kao [Memorials of Chang Chih-tung] . Vol. 17, n.p., 1920.

Chang Hsin-ch'eng. Chung-kuo hsien-tai chiao-tung shih [A History of China's modern communications] . Shanghai: Friendship Press, 1931.

Chang Kia-ngau. Chung-kuo t'ieh-tao chien-she [China's struggle for railroad development] Shanghai: Commercial Press, 1947.

________. China's Struggle for Railroad Development. New York: John Day Company, 1943.

Chang K'oh-wei. Tung-pei kang jih ti tieh-lu cheng-tse [Railway policy for the resistance against Japanese in the North-east] . n.p., 1913.

Chang Pe. Jih-pen tui hua chih chiao-tung chin-lueh [The Japanese agression to Chinese communication] . Shanghai: Commercial Press, 1931.

Chao Tseng-chiao. Chan-ho ti chiao-tung chien she [Post war reconstruction of communications] . Shanghai: Commercial Press, 1947.

Chao Yung-hsin. Chung-kung ti tieh-tao kung-tso [Railway works in Communist China] . Hong Kong: Union Press, 1954.

Ch'en Chin-sung. "The Planning of a National Railway System for China". M.C.E. Thesis, Cornell University, 1928.

Ch'en Chio. Tung-pei lu kang hsin-lin wen-t'i [Railway, mining, and forest problems in Northeastern China]. Shanghai: Commercial Press, 1933.

Ch'en Hui. Kuanghsi chiao-t'ung wen-t'i [Transport and communications in Kuanghsi province]. Ch'ang-sha: Commercial Press, 1938.

________. Chung-kuo t'ieh-lu wen-t'i [China's railway problems].. Peking: San-luan Books, 1955.

Ch'en Po-chuang and Huang Yin-lai. Chung-kuo hai-kuan t'ieh-lu chu-yao hang-p'in liu-t'ung kai-k'uang [Statistics of commodity flow of Chinese maritime customs and railways, 1912-1936]. Shanghai: Chiao-t'ung University, 1937.

Cheng Lin. The Chinese Railways, a Historical Survey. Shanghai: China United Press, 1935.

________. The Chinese Railways - Past and Present. Shanghai: China United Press, 1935.

Ch'i Yu. Hsin-chung-kuo ti t'ieh-tao chien-she [New China's railway construction]. Peking: San-luan Books, 1953.

Chiao-chi t'ieh-lu Kuan-li Chu. Min-kuo erh-shih-san (and ssu) nien-tu ts'ai-liao t'ung-chi pao-kao. [Statistical material on the Chi'ingtao to Chinan Railway, 1934-35]. Chinan: n.p., 1935.

Chiao-tung lei-pien chia chi ,[Classified compendium on communications, first collection]. 3 vols. Peking: Chiao-t'ung ts'ung-pao she, 1918.

Chiao-tung shih lu-cheng pien [History of communications: Road and railway administration]. Vol. 1, Nanking, n.p., 1935.

Chin Chia-fung. Chung-kuo chiao-t'ung chih fa-chan chi ch'i Ch'u-hsiang [Transportation development and trends in China]. Shanghai: Cheng-chung shu-chu, 1937.

Chin Shih-hsuan, ed. Chung-kuo tung-pei t'ieh-lu wen t'i hui-lün [China's Northeastern railway problems]. Tientsin: Ta Kung Pao, 1932.

________. T'ieh-lu yu k'ang-chan chi chien-she [Railways, resistance war, and reconstruction]. Shanghai: Commercial Press, 1947.

________. Tieh-lu yun-shu ching-yen tan [Experiences on railway transportation]. Chungking: Cheng Chung, 1943.

________. T'ieh-tao yun-shu hsueh [Railway transportation]. Shanghai: Commercial Press, 1948.

Chin, Steve S. K. Mao Tse-tung shih-hsiang, shing-shih yu nei-yung [The thought of Mao Tse-tung: form and content]. Hong Kong: University of Hong Kong Press, 1976.

China. Chiao-t'ung kuan-pao [Communications Gazette]. Peking: Ministry of Posts and Communications, Translation Bureau Series Monthly, 1909-1911.

China (PRC). China's Railways: A Story of Heroic Reconstruction. Peking: Foreign Language Press, c. 1951.

________. Tieh-lu huo-wu ch'i-nan [Railway freight guide]. Shanghai: Shanghai Railway Bureau, 1951.

________. Jen-min chiao-tung [People's Communications]. Peking: Chung-yang jen-min cheng-fu chiao-tung pu, June 1950 - Semi-monthly. (Monthly prior to 1955.)

________. Foreign Trade in Machinery and Transportation Equipment Since 1952, U.S., C.I.A. Research Aid, A(ER) 75-60. Washington, D.C., January, 1975.

________. Ch'uan-kuo t'ieh-lu lu-k'o lieh-ch'e shih-k'e-piao, 1960 nien hsia-ch'i shih-hsing (Translated into English by JPRS, no. 8153: Communist China Railroad Passenger Timetable, Summer 1960). Peking: Peoples' Railway Publishing House, 1960.

________. Chung-hua-jen-min-kung-ho-kuo fa-chan kuo-min ching-chi ti ti-i-ko wu-nien-chi-hua [The First Five-Year Plan of the Peoples' Republic of China]. Peking: Peoples' Publishing House, 1955.

________. Chung-kuo-kung-chan-tang ti-pa-ch'i chuan-kuo tai-piao ta-hui . . . ti ti-erh-ko wu-nien chi-hua . . . [The Second Five-Year Plan of the Peoples' Republic of China . . .]. Peking: n.p., 1956.

________. State Statistics Bureau. Wei ta ti shih-nien [The great ten years]. Peking: Peoples' Press, 1959.

________. Ch'uan-kuo t'ieh-lu lu-k'o lieh-ch'e shih-k'e piao, July-Nov., 1963 (Translated into English by JPRS, no. 21963: Communist China Railroad Passenger Timetable, July-November 1963). Peking: Railway Publishing House, 1963.

________. Ministry of Railways. Organizing Railway Transportation: Communist China (English translation of T'ieh-lu yun-shu kung-tso tsu-chih). Washington, D.C.: U.S. Government Printing Office, 1965.

________. Ministry of Railways, Sha-mo ti ch'u-lu ching-yen. Peking: February 1960. (Translated by JPRS, no. 11770: Railroad Construction Experience in Desert Areas, 1961).

________. Ministry of Railways. Chuan-kuo ho-ch'e shing-ch'e shih-k'e-piao. Peking: Ministry of Railways, 1956.

________. Ministry of Railways. Huo-wu Yun-chia lu piao [Freight tariff tables]. Peking: People's Railways Press, 1955.

________. Ministry of Railways. Huo-wu Yun-chia Chi-suan kuei-tze huo-wu p'in-ming fen-lei piao [Rules for computing freight rates and freight classification tables], Peking: n.p., 1955.

China (PRC). The People's Liberation Army. *Chung-kuo jen-min chieh-fang chan-cheng san-nien chan-chi* [Tabulations of the campaigns of the PLA, 1946-1949] n.p., Chung-kuo jen-min chieh-fang-chun tsung-pu, 1949.

China (Republic). *Ch'uan-kuo chiao-t'ung hui-i*. Ch'uan-kuo chiao-t'ung hui-i hui-pien [Reports of the National Transport Conference, compiled by the Chiao-t'ung pu]. Shanghai: Shang-wu yin-shu-kuan, 1928.

________. Bureau of Geological Survey. *Chuan-kuang tieh-lu ch'u-kan pao-gao* [Primary reports on Ssuchuan-Kuangyuan Railway]. Shanghai: Geological Survey, 1931.

________. Ministry of Railways. *Report on Railway Finance*. Shanghai, *Mimeographed*, 10 December 1929.

________. Ministry of Communications. *Railways in Communist China*. Taipei: Communications Research Bureau, 1961.

________. Ministry of Finance. *Kuo-yo T'ieh-lu tsai-tan fen-lei chung-piao* [Classified railway loans of Chinese National Railways]. Unpublished, 1936.

________. Ministry of Railways. *Chung-hua min-kuo t'ieh-lu pu-tung huo-wu fen-t'eng piao*. [Classification of general railway goods of the Republic of China].

________. *Railway Loan Agreements of China*. Nanking: Ministry of Railways, 1937.

________. Ministries of Communications and Railways. *Chiao-tung Shih, lu-cheng pien*. [History of communications: road administration]. 37 vols. n.p., 1930-1937, 23,776 pp.

________. Ministry of Communications. *Chinese Government Railways*. Peking: n.p., 1920.

________. Communications Delegation to Europe and America, ed. *Kao-cha o-mei chiao-tung pao-gao* [Report of the European-American Communications Study Tour]. Shanghai: Commercial Press, 1935.

________. Ministry of Communications. *. . . Minutes of the Standing Committee on the Unification of Railway Accounts and Statistics*. n.p., 1925.

________. (written by Chang Yi-yao). *Fei-chu kung-lu she-shih kai-fang* [Highway development in Communist China]. Taipei: Communications Research Institute, 1958.

________. General Post Office. *Chung-kuo tung-yiao t'i-fang wu-ch'an chi* [Products of China's postal areas]. Shanghai: Commercial Press, 1937.

________. Ministry of Railways. *Chung-kuo t'ieh-tao pien lan* [A convenient guide to Chinese railways]. Shanghai: Commercial Press, 1934.

China's Railway. Peking: Foreign Language Press, 1950.

China Reconstructs 14 (July 1965).

Chinese Eastern Railway Company. North Manchuria and the Chinese Eastern Railway. Harbin: C.E.R. Printing, 1924.

Chiu Kuang-fu. Chung-kuo t'ieh-lu chi-ch'i fa-chan ti chu-hsiang [China's railway and its development trends]. Hong Kong: Freedom Press, 1950.

Chou En-lai. Chung-kuo-kung-chan-tang ti-pa-ch'i ch'uan-kuo tai-piao ta-hui . . . ti ti-erh-ko wu-nien chi-hua . . . [The Second Five-Year Plan of the People's Republic of China . . .]. Peking: n.p., 1956.

Chou Shan-p'ei. Hsin-hai Ssu-ch'uan cheng-lu ch'in-li chi [A record of my personal experiences in the Ssuch'uan railway strife of 1911]. Chungking: Ch'ung-ch'ing jen-min ch'u-pan she, 1957.

Chung-hua-jen-min-kung-ho-kuo fa-chan kuo-min ching-chi ti ti-i ko wu-nien chi-hua [The First Five-Year Plan of the People's Republic of China]. Peking: n.p., 1955.

Fang Ch'iu-wei. Chung-kuo pien-chiang wen-ti shih-chiang [Ten lectures on China's frontier problems]. Shanghai: Motor Press, 1937.

Fei Hsiao-tung, Hsiang-t'u chung-kuo [Earthbound China]. Hong Kong: Feng-huang Press, n.d., reprinted from 1947 version.

Fu Yu-cheng. Chung-kuo chin-t'ai tieh-lu shi chi-liao, 1863-1911, 3 vols. [Modern railway history materials of China, 1863-1911]. Peking: China Books, 1963.

Griffin, Eldon. China's Railways as a Market for Pacific North-west Products. Seattle: University of Washington, 1946.

Hammond, F.D. Report on the National Railways of the Republic of China. London: n.p., 1935.

Han To-fu, et al. Hsin-chung-kuo ti chiao-tung yun-shu shih-yeh [New China's communication and transportation enterprise]. Peking: San-luan Books, 1953.

________. Kuo-tu shih-ch'i ti chiao-tung yun-shu [Communication and transportation in a transition period]. Peking: People's Press, 1955.

________. Chiao-tung yun-shu shih-yeh yu kuo-chia kung-yeh hua [Communication and transportation enterprises and national industrialization]. Peking: Workers' Press, 1955.

Herold, Edward. Railway Enterprise in China. London: n.p., 1907.

Hsieh Pin. Chung-kuo t'ieh-tao shih [History of Chinese railways]. n.p., China Press, 1929.

Hsiung Heng-ling, and Ho Te-chuan. Kuo-fu hsueh-shuo yu chung-kuo t'ieh-lu [Dr. Sun Yat-sen's Doctrine and Chinese railways]. 2 vols. Taipei: Railway Nationalist Party Committee, 1966.

Hsu Ch'i-heng. Chan T'ien-yu ho chung-kuo t'ieh-lu. [Chan T'ien-yu and China's railways]. Shanghai: Jen-min, 1957.

Hsu, Mongton Chih. "Railway Problems in China." New York: Ph.D. Dissertation, Columbia University, 1915.

Hsu Yueh-sheng."Economic Construction of Railroads for China". Master in Civil Engineering Thesis, Cornell University, 1912.

Hsueh Kuang-chien. Chiao-tung hsing-ching yen-ch'iu [A study of communication administration]. Chungking: Commercial Press, 1944.

Huang, Peter Pao-chao."Methods of Railway Development and Location in the United States, and Their Application to China".Master in Civil Engineering Thesis, Cornell University, 1922.

Huang, Po-chiao. I ko wu-nien chien chih ching-hu, hu-hang jung tieh-lu chung-wu shing-cheng [An administrator-general of the Peking-Shanghai and Shanghai-Hangchou-Ningpo railways for five years] n.p., 1940.

Hung I. Kung-fei chiao-tung chien-she shih yen-ch'iu [A study of communication development in Communist China]. Taipei: Yang-ming Shan-chuang, 1957.

I Yeh. Kuei-Chien lu shang tsa i [Memoirs on the Kuei-Chien Railway]. Hong Kong: Chih Cheng Press, 1971.

Ilyin, A. I., and Voronicher, M. P. Railway ransport of the Chinese People's Republic. Moscow: n.p., 1959. (Translated from Russian in JPRS, no. 3484).

Kann, E. The History and Financing of China's Railways. Shanghai: n.p., 1937.

Kennan, George. E. H. Harriman's Far Eastern Plans. New York: Country Life Press, 1917.

Kent, Percy H. B. Railway Enterprise in China. London: Edward Arnold, 1907. Translated into Chinese by Li, Pao-heng. Chung-kuo t'ieh-lu fa-chan shih. Peking: San-luan Books, 1958.

Kinder, C. W. Railways and Collieries in North China. Institution of Civil Engineers, n.d.

Kingman, Harry L. Effects of Chinese Nationalism Upon Manchurian Railway Developments, 1925-1931. Berkeley: University of California Press, 1932.

Ko Sui-ch'eng. Chung-kuo chi chiao-tung [China's communications]. Shanghai: China Press (1927), 1940, 3rd. edition.

Kung Hsueh-sui. Chung-kuo chan-shih chiao-tung shih [A history of China's war-time communications]. Shanghai: Commercial Press, 1947.

Li Hung-chang. Li wen-chung-kung ch'uan-chi [Complete works of Li Hung-chang]. n.p., n.d.

________, ed. Chi-fu tung-chih [Imperial records of Chihli province] n.p.: 301 Chuan, 1884. Shanghai: reprinted, Commercial Press, 1934.

Li Kuo-ch'i. Chung-kuo chao-ch'i ti t'ieh-lu ching-ying [Early railway development in China]. Taipei: Academia Sinica, 1961.

Ling Chuke. China's Railway Rolling Stock. Seattle: University of Washington Press, 1946.

Ling Hung-hsun. Chan T'ien-yu hsien-sheng nien-p'u [Data for biography of the railway engineer]. Taipei: Chung-Kuo Kung-ch'eng-shih-hsueh-hui, 1961.

________. Shih-liao nien chu-lu sheng-yeh [Sixteen years of railway building life]. Taipei: Biographic Literature Press, 1968.

________. Chung-kuo t'ieh-lu chih [Chinese railways]. Taipei: Hsieh Lung Printing, 1954.

________. Chung-kuo t'ieh-lu kai lun. [On Chinese railways]. Taipei: National Publishing Bureau, 1950.

________. Chih-shih nien-lai tung-ch'ing, chung-tung, chung-chang t'ieh-lu pien-chien chi ching-kuo [Seventy years of the Chinese Eastern and South Manchurian railways]. Taipei: Ministry of Communications, 1965.

Liu Ch'ai-hsing. Wo-kuo ti yun-shu ti-li [Our country's transportation geography]. Peking: Science Popularization Press, 1957.

Liu Chuang. Wo-kuo ti yun-shu-yeh [China's transportation industry]. Peking: China Youth Press, 1956.

Liu K'e-shu. Chung-kuo chiao-tung [China's communications]. 2 vols. Taipei: Chinese Culture Publishing Enterprise Committee, 1954.

Liu Ming-chuan. Liu chuang shu kung tso yi [Memorials by Liu Ming-chuan]. n.p., 1944.

Mao Tse-tung. Mao Tse-tung shih-hsiang wan-shui [Long live the thought of Mao Tse-tung]. n.p., 1969.

Mo Chien-tseng and Li Ying-chao. Ch'uan-kuo t'ieh-lu kuan-li chih-tu [The system of managing the national railways]. Peiping: Chiao-t'ung University, 1936.

North China Herald, 1884, p. 700.

Pai Shou-i. Chung-kuo chiao-tung shih [A history of China's communications. Taipei: Commercial Press, 1965.

Parsons, William Barclay. An American Engineer in China. New York: McClure, Phillips & Co., 1900.

Peking Review, April 5, 1960, p. 11.

Peking-Shenyang Railway. Pei-ning t'ieh-lu yen-hsien ching-chi tiao-ch'a pao-kao shu [Report of the investigation of economic conditions along the Peking-Shenyang Railway]. n.p., 1937.

P'eng Yun-o. Ti-i-ko wu-nien chi-hua chung ti t'ieh-lu chien-she [Railway construction during the First-Five-Year plan] Peking, n.p., 1956.

Pi I-tsun. Chuan shih hsi-nan tien-ti chien [The retreat to the Southwest]. Hong Kong: Chi shih Nien-tai, 1973.

P'ing-han T'ieh-lu Kuan-li Wei-yuan-hui. P'ing-han t'ieh-lu nien chien [Yearbook of the Peiping-Hankow Railway]. Hankow: n.p., 1932.

Rosenbaum, Arthur L. "China's First Railway, The Imperial Railways of North China, 1880-1911." Ph.D. Dissertation, Yale University, 1971.

Shen Tsou-t'ing. T'ieh-lu wen-t'i tao-lun chi [Essays on railway problems]. Shanghai: Commercial Press, n.d.

________, et al. T'ieh lu huo-yun yeh wu [On railway transport]. Shanghai: Commercial Press, 1935.

Sheng Chu-kung. Chiao-tung ti-li [Transportation geography]. Shanghai: Commercial Press, 1931.

Sokolsky, George. The Story of the Chinese Eastern Railway. Shanghai: North China Daily News & Herald Ltd., 1929.

Stringer, H. The Chinese Railway System. Tientsin: Tientsin Press, 1925.

Su-hang-yung t'ieh-lu tang. [Document concerning the Shanghai-Hangchou-Ningpo Railway] n.p., n.d.

Sun, E-tu (Zen). Chinese Railways and British Interests, 1898-1911. New York, King's Crown Press

Sun Yat Sen. San Min Chu I. Shanghai: n.p., 1929.

________. Chien kuo fong lueh. Taipei: Chengchung Press, 1952. (reprinted).

South Manchuria Railway Company. Report on Progress in Manchuria. Dairen: S.M.R., 1929, 1931, 1932, 1934, and 1936.

________. Manchuria: Land Opportunities. New York: Thomas F. Logan, 1922.

Tai Chih-li, ed. Ssuch'uan pao-lu yun-tung shih-liao [Historical materials on Ssuch'uan Railway Preservation Movement]. Peking, 1959.

Ten Great Years. Peking: Foreign Language Press, 1960.

T'ieh-tao k'o-hsueh chi-shu [Railway technique]. Monthly. Peking: Jen-min t'ieh-tao ch'u-pan-she, 1959.

T'ieh-tao She-hui [Railway Union], ed. Min-kuo t'ieh-lu i-nien shih [Railways during the first year of the Republic]. Peking: Chinghua, 1914.

Ting Ming-nan, et al. Ti-kuo chu-i chin hua shih [History of imperialist invasion of China]. Peking: Jen-min chu-pan-she, 1961.

Treaties and Agreements Between China & Japan Concerning the Railways of the Three Eastern Provinces. 2 vols. Taipei: Ti-tao Pu, 1969.

Tseng Chung-ming. Lu-cheng lun-ts'ung. Shanghai: K'ai-ming shu tien, 1954.

Tseng K'un-hua. Chung-kuo t'ieh-lu hsien-shi tung-lun [On China's current railway situation]. Hua-hua Railway Institute, n.p., 1907.

________. Chung-kuo t'ieh-lu shih [A History of Chinese railways]. Peking: n.p., 1924.

Tuan I-ming. K'ung-fei chiao-tung chi yen-ch'iu [A study of communications in Communist China]. Taipei: Institute of National Defence, 1960.

U.S. Strategic Intelligence Branch, Military Intelligence Division, U.S. Army. Railroads of China. vols. 1-7. Washington D.C.: Strategic Engineering Study, S.E.S. no. 145, November 1944.

Volpicelli, Z. (pseud. Vladimir). Russia on the Pacific and the Siberian Railway. London: Sampson Low, Marston & Co., 1899.

Von Lochow, H. J. China's National Railways. Peiping: National Peking University, 1948.

Wang Ching-ch'un, et al. Chung-kuo t'ieh-lu chieh k'uan ho tung ch'uan chi [A complete collection of China's railway loans]. Taipei: Student Press, 1922.

Wang Chin-yu. Ch'in-tai chung-kuo ti tao-lu chien-she [Modern China's railway and road construction] Hong Kong: Lung-mun Books, 1969.

Wang Cho. Chiao-tung shih [A history of communications]. Shanghai: Commercial Press, 1932.

Wang Kuang. Chung-kuo chiao-tung k'ai-lun [On China's communications]. Taipei: Chengchung, 1953.

________. Chiao-tung k'u ch'in tan [Communications] Taipei: Ocean Transportation Press, 1956.

________. Kai kuo liu-shih nien chiao-tung shih lun [Sixty years of communications - History of the Republic]. Taipei: By the Author, 1971.

Wang Yu, et al. Ch'u-fang Huang-sha Ho. Chungking: People's Publishing House, January 1957. (Translated into English by JPRS no. 16933, 1963: Visit to Huang-sha River and Ch'eng-tu to Pao-chi Railroad Construction).

Williams, E. T., trans. Recent Chinese Legislation Relating to Commercial, Railway, and Mining Enterprise. Shanghai: Shanghai Mercury, 1904.

Witte, Sergey Yulyerich, Count. Memoirs of Count Witte. Translated by A. Yarmotinsky. n.p.: Doubleday, Page, and Co., 1921.

Wu Ying-hua. Erh-shih nien lai ti nan-man-chou t'ieh-tao chu-shih hui-she [Twenty years of the South Manchurian Railway Company Limited]. Shanghai: Commercial Press, 1930.

Wu, Yuan-li. China's Transportation "The Adequacy of Domestic Supply, Trade and Transportation of Coal in Mainland China". (Interim Report No. 3), Stanford Research Institute, June 1954.

________. The Spatial Economy of Communist China. New York: Frederick A. Praeger, 1967.

Yang-cheng wan-pao [Canton Evening Post]. 10 October 1964.

Yang Hsiang-nien. T'ieh-tao ching-chi yu tsai-cheng [Railway economics and finance]. Chung-king: Commercial Press, 1944.

Yang Tu. Yueh-han t'ieh-lu i [A discussion of the Hankow-Canton Railway]. n.p., n.d.

Yeh Kung-ch'o. Hsia-an hui-kao [Collected poetry and correspondence relating to railways]. n.p., 1930.

Young, Walter. Japanese Jurisdiction in the South Manchuria Railway Areas. Baltimore: n.p., 1931.

Yu Fei-peng. Shih-wu-nien lai chi chiao-tung k'ai-fang [Communications in the last fifteen years]. n.p., 1946.

B. Articles

Altree, Wayne. "A Half-Century of the Administration of the State Railways of China." Papers on China 3 (1949): 78-133.

Anderson, G. E. "Railway Situation in China." U.S. Special Consular Report, no. 48 Washington, D.C., 1911, p. 25.

Chang Ching-chih. "Nothing Stops the Railway Builders." In China in Transition, Peking: China Reconstructs, 1957, p. 171.

Chang Kuo-wu and Chang Chih. "An Examination of the Development, Location, Support to Agriculture of Local Railways in Our Country." Ti-Li [Geography], no. 1, 1964, pp. 5-10.

Chang Li. "Railway Construction in China," Far Eastern Survey, 25 March 1953, pp. 37-42.

Chang Wu-tung and Wu Cho-liang. "On the Location of Transportation Networks in Regional Planning." Ti-Li [Geography], September 1961, pp. 216-220.

Chao Sheng-chuan and Pan Pei-wen. "The Threat of Windblown Sand to Railways and Its Prevention." Ti-Li [Geography], no. 1, 1965, pp. 13-17.

Chen Han-en. "An Analysis of the Relationship between Transportation and the Location of Mining Industries." Ti-Li [Geography], no. 5, 1964, pp. 207-210.

Chinchou Railway Administrative Bureau, CCP Committee of the, "Continuously Raise the Quality of Equipment: Realize the Potential of Rail Transport," Red Flag, no. 22 (16 November 1960), pp. 41-45.

"Chinese Owned Railroads in Manchuria." Far Eastern Review, November 1926.

Chou Shun-hsin. "Railway Development and Economic Growth in Manchuria." China Quarterly, 45 (1971): 57-84.

Ch'uan Han-sheng. "Ch'ing-chi t'ieh-lu chien-she ti tzu-pen wen-t'i." ["The problem of financing railway construction in the late Ching period"] in Journal of Social Science (Taipei. Taiwan University), 1 (1953): 1-16.

________. "T'ieh-lu kuo-yu wen-t'i yu Hsin-hai Ke-ming," [The nationalization of railways and the revolution of 1911] in Chung-kuo hsien-t'ai shih chung-kan (Taipei. Chung-cheng), 1 (1960): 209-270.

Chung K. F. "A Geographical Study of the Railway Route between Lanchou and Tahcheng." Ti-Li [Geography], July 1944, pp. 71-80.

Currie, Blair. "The Woosung Railroad, 1872-1877." Papers on China 20 (1966): 49-85.

Davis, M. W. "Railway Strategy in Manchuria." Foreign Affairs, April 1926, p. 499.

Denby, C. and Allen, E. P. "Chinese Railway Development," Engineering Magazine (London), 16 (1898): 339-348.

"Economic Bases for New Railways in Manchuria." Far Eastern Review, May 1927.

"Foreign Supervision over China's Railways," Far Eastern Review, September 1923.

Freyn, Hubert. "The Kowloon-Hankow Railway and Japanese Air Raids." China Journal 29, (November 1938): 257 pp.

Ginsburg, Norton S. "China's Railroad Network," Geographical Review 41 (1951): 470-494.

________. "Manchurian Railway Development," Far Eastern Quarterly 8 (August 1949): 398-411.

Hsin-min wan-pao ["New People's Evening Post"], 11 April 1965, Shanghai.

Irick, Robert. "The Chinchou-Aigun Railroad and the Knox Neutralisation Plan in Ch'ing Diplomacy." Papers on China 13 (1960): 8-112.

Joint Publication Research Service. "Success of Railroad Transport in China." JPRS no. 4296 (1961). Translated from Zheleznedorozhnyy Transport [Railroad Transport], Moscow, 1960, no. 6, pp. 81-83.

________. Communist China Railroad Passenger Timetable, July-November 1963, JPRS no. 21963, U.S. Department of Commerce, Washington, D.C., 1963.

_________. "Local Railways in Communist China." JPRS no.5710 (1960). Translated from Kung-jen jih-pao, Peking, 20 August 1960.

_________. "Hsining to Shanghai Summer Train Schedule, April 1960." JPRS no. 6479 (1960). Translated from Ch'inghai Jih Pao, 21 April 1960, p. 4.

_________. "New Successes of the Railroad in the Peoples' Republic of China." JPRS no. 10962 (1961). From Zheleznedorozhnyy Transport [Railroad Transport], 1960, no. 7.

_________. "Development of Transportation Network Should Be Related to Development of Agriculture and Industry." JPRS no. 23113 (1964). From People's Daily, 3 December 1963.

_________. "Construction of Bridges and Roads in Chechiang Province." JPRS no. 18625. (9 April 1963). (original source lost).

_________. "China's Transportation, 1959-1960." JPRS no. 3609 (1959-60). (from a number of Chinese sources too numerous to be named).

_________. "A Study of the Topography of a Desert Region for the Selection of a Railway Route." JPRS no. 33258 (1965). From Ti-Li, Peking, 4 27 July 1965, pp. 166-170.

_________. "An Analysis of the Relationship Between Transportation and Organization of Extractive Industries." JPRS no. 29013 (1964). From Ti-Li [Geography], Peking, no. 5 29 September 1964, pp. 207-210.

_________. "Organizing Railway Transportation." JPRS no. 28854 (1964). From T'ieh-lu Yun-shu Kung-tso Tzu-chih, Peking: People's Railway Publishing House, June 1964, pp. 1-495.

_________. "Railways Form the Artery of China's National Economy." JPRS no. 25062 (1964), from People's Daily, Peking, May 12, 1964, p. 4.

_________. "The Development of the Communist Chinese Railroad System." JPRS no. 18093 (1962). From Deutsche Eisenbahn Technik [German railroad technology], East Berlin, no. 12 December 1962, pp. 597-599.

_________. "Fully Develop the Pivotal Function of Communication and Transportation." JPRS no. 22927 (1964). From Jen-min Jih-pao, Peking, December 3, 1963, p.5.

_________. "Transportation System in China." JPRS no. 4484 (1961). From Cahiers France-Chinois [Franco-Chinese friendship]. Paris, no. 7 (September 1960), pp. 42-52.

Lei T'ing and Liang K'uang-pai. "The Role of Transportation in Our National Economy." Ching-chi yen-chiu [Economic study], no. 2, February 1965, pp. 39-43.

Leung Chi-keung. "Railway Building and National Goals: Some Reflections on Chinese Development Strategies." Contemporary China Seminar, Centre of Asian Studies, University of Hong Kong, 1976.

Leung Chi-keung. "Transportation and Spatial Integration." In China: Development and Challenge, 4 vols. Edited by Lee Ngok and Leung Chi-keung. Proceedings of the Fifth Leverhulme Conference, Centre of Asian Studies, University of Hong Kong, 1979.

________. "Regional Inequality and Chinese Development Strategy." Paper Presented at the Institute of British Geographers Annual Conference, Manchester, January 1979.

________. "A Review of the Thought of Mao Tse-tung: Form and Content by Steve S. K. Chin," The China Quarterly, no. 68, (1976), pp. 845-848.

Li Chien-ch'ao. "Wo-kuo yu i-t'iao tien-chi-hua tieh-lu - Yang-An tieh-lu" [Another electric railroad of our country - Yang-An Railway]. Ti-li chih-shih, no. 7, 1978, pp. 1-3.

Li Fu-ch'un. "Report on the Draft National Economic Plan for 1959." Hsin-hua pan-yueh-k'an [New China Semimonthly], no. 9, Peking: n.p., 1959.

________. "Report on the Draft of the National Economic Plan for 1960." Jen-min jih-pao, March 31, 1960.

Li Hsien-nien, "Report on the Actual Results of the 1958 State Budget and the Plan for 1959 State Budget." Hsin-hua pan-yueh-k'an [New China Semimonthly], no. 9, Peking: n.p., 1959.

________. "Report on the Actual Results of the 1959 State Budget and the Plan for the 1960 State Budget." Jen-min jih-pao, April 1, 1960.

________. "Report on Finance to the National People's Congress." People's Handbook, 1960, p. 185.

Li Hui-min. "The Ssuchuan-Kueichou Railway," Chinese Communist Affairs, vol. 3, no. 2, (April 1966) pp. 46-49, map.

Ling Hung-hsun. "A Decade of Railroad Construction, 1926-36." In The Strenuous Decade: China's Nation-Building Efforts, 1927-1937. Edited by Paul K. T. Shih. New York: St. John's University Press, 1970.

Lu Cheng-ch'ao. "Tsen-yang wan-ch'eng chin-nien ti t'ieh-lu yun-shu jen-wu" [How to fulfill this year's railway transportation quota], Hsin-kua pan-yueh-k'an [New China Semi-monthly], Peking, no. 10 (May 1959), pp. 32-33.

________ "Wo-kuo tieh-lu chien-she ti tao-lu," [The road to our country's railway construction], Hungch'i, no. 2 (1959), p. 16.

Munthe-Kaas, H. "Roads and Rails in China," Far Eastern Economic Review, Hong Kong, 17 February 1966, p. 326.

Murphey, Rhoads. "China's Transport Problems and Communist Planning," Economic Geography 32 (1956): 17-28.

Petrov, Victor P. "New Railway Links between China and the Soviet Union," Geographical Journal, 122 (1956): 471-477.

Phillips, Don. "We Don't Have a Locomotive Named for Nixon, but They Have One Named for Mao Tse-tung," Trains Magazine, November 1972, pp. 36-49.

Po I-po. "Kuan-yu chiao-tung yun-shu ho yiao-tien kung-tso ti fa-yen," [A speech on the work of communications, transportation, and postal communications] in Wu-nien lai ti ch'ai-cheng ch'ing-chi k'ung-tso [Five years of financial and economic work]. Peking: Financial and Economic Press, 1955.

"Policy and Function of Railway Freight Rate Determination," Jenmin Tiehtao. [People's Railway], no. 9, 1952.

Prybyla, Jan S. "Transportation in Communist China." Land Economics, vol. 42, no. 3, 1966, pp. 268-281.

"Railway Politics in Manchuria." China Weekly Review, April, 1927.

Ren-min Shou-ts'e, [People's Handbook], 1951, p. 214.

Rigby, Edward, and Leitch, William. "Railway Construction in North China," Proceedings of the Institution of Civil Engineers 160 (1905): 272-314.

Royama, M. "The South Manchuria Railway Zone." Pacific Affairs. 3 (November, 1930): 1018-1034.

Shaw, Arthur M. "Transport Trends in China." Quarterly Review of Chinese Railways 1 (1 January 1937): 135-146.

"Speed Up the Modernization of Transportation and Communications So As To Realize the Will of Chairman Mao." People's Daily, 6 November 1977.

Sun Fo. "National Scheme of Railway Construction." China Year Book. Tientsin: n.p., 1929-1930.

Sun Zen E-tu. "The Shanghai-Hangchou-Ningpo Railway Loan of 1908." Far Eastern Quarterly, vol. 10, no. 2, (February 1951).

________. "The Pattern of Railway Development in China." Far Eastern Quarterly 14 (February 1955): 179-199.

Ta Kung Pao, Hong Kong, 8 June, 24 July, 1976; 1 October, 18 November, 1977; 13 June, 1978.

Tan, P. L. "Traffic and Operation, Traffic Conditions of the Chinese National Railways." Quarterly Review of Chinese Railways 1 (1 July 1936): 57-68.

Teng Tai-yuan. "The People's Railways Since Liberation." In New China's Economic Achievements 1949-1952. Compiled by China Committee for the Promotion of International Trade. Peking: Foreign Language Press, 1952, pp. 209-217.

Wang C. C. "The Chinese Eastern Railway." Annals American Academy of Political and Social Science 122 (November 1925): 57-59.

Wang C. C. "The Administration of the Chinese Government Railways." Chinese Social and Political Science Review. 1 (1916): 68-85.

Wang Shou. "The Transport and Communications Industry in the Continuous Big Leap Forward." Jen-min jih-pao [People's Daily], Peking, 19 February 1960, p. 7.

Wang Shou-tao. "Develop Transport, Postal and Telecommunication Undertakings." New China Fortnightly no. 88 (21 July, 1956), p. 133.

________. "Continuously Develop Communications and Transport Undertakings: Improve Service to Production and Livelihood." Jenmin Jihpao [People's Daily], 26 May 1961, p. 7.

Wang Yeh-chien. "Chia-wu chan-cheng i-ch'ien ti chung-kuo t'ieh-lu shih yeh." [Railway enterprise in China prior to the Sino-Japanese War of 1895] in Chung-yang yen-chiao yuan, li-shih yu-yen yen-chiu sho chi-kan. Nanking: 31 (December 1960): 167-189.

Wen Liang. "General Conditions in Railroad Transportation on the Mainland." China Weekly. Hong Kong, 11 March 1916, pp. 6-8, 11.

Wu To. "Chin-t'ung t'ieh-lu ti cheng-i." [The debate over the Tientsin-Tungchou Railway]. Reprinted in Chung-Kuo Chin-t'ai shih lun-chung, Series I, vol. 5. Taipei: Cheng Chung, 1956. pp. 135-170.

Yang Ch'ing-K'un. "Chung-kuo hsien-tai k'ung-chien chu li chih shu tuan." [The construction of space in modern China]. Ling-nan hsueh pao 10 (December 1949): 154.

Yang, Wu-yang. "An Assessment and Regional Demarcation of the Natural Conditions for Land Transportation in China." Ti-Li hsueh-pao [Acta Geographica Sinica] 30 (4) (December 1964): 301-318.

Yin A. I. "Railroad Transport in the Chinese People's Republic," JPRS no. 3484, 6 July 1960.

Yu I. Formin. "Success of Railroad Transport in China." Zheleznedorozhnyy Transport [Railroad transport], Moscow, no. 6. See also JPRS., no. 4296, 29 December 1961.

C. Yearbooks & Statistical Sources

Aird, John S. Population Estimates for the Provinces of the People's Republic of China, 1953-1974. U.S. Bureau of the Census, 1974.

Chen, Nai-ruenn. Chinese Economic Statistics. Chicago: Aldine Publishing Company, 1967.

Chiao-t'ung-pu t'ung-chi pan-nien pao. [Semi-annual statistics issued by the Ministry of Communications]. January-June, 1934. Republic of China. Nanking: Chiao-t'ung-pu 1935.

China (PRC) Handbook of Economic Indicators. C.I.A. Research Aid, ER 76-10540, August, 1976.

China Handbook. New York: Macmillan Co., 1937-1945, and Taipei: China Publishing Company, 1952-present. (Retitled China Yearbook after 1957/58 issue).

China Industrial Handbook. Republic of China, Ministry of Industry. Shanghai: 1933-1936.

China Year Book. Shanghai: North China Daily News and Herald, 1912-1939.

Chinese Economic and Statistical Review, The. 8 vols. Shanghai: China Institute of Economic and Statistical Research, 1934-1941.

Chinese Year Book, The. [Chung-kuo Nien-chien]. Chungking: Commercial Press, and Bombay: Thacker & Co., 1935/36 to c. mid-1940's.

Chung-hua kuo-yo tieh-lu shing-che tung-chi yueh-kan [Train statistics of Chinese National Railways Monthly]. Republic of China, Ministry of Railways. July-December, 1932, and January-June, 1935.

Chung-hua kuo-yo tieh-lu tung-chi yueh-kan [Railway statistics, Chinese National Railways Monthly]. Republic of China, Ministry of Railways. 1933, 1934.

Chung-hua-min-kuo t'ung chi t'i yao. [Statistical abstract of the Republic of China]. Republic of China, 1935.

Chung-kuo ching-chi nien-chien [The Chinese economic yearbook]. Nanking: Commercial Press, 1932-36.

Chung-kuo ching-chi nien-chien [The Chinese economic yearbook]. Hong Kong: Pacific Economic Research Institute, 1947, 1948.

Chung-kuo jen-min chieh-fang chan-cheng san-nien chan-chi [Tabulations of the campaigns of the People's Liberation Army, 1946-1949]. n.p., 1949.

Clark, Grover. Statistical Data on China. New York: American Council, Institute of Pacific Relations, 1932.

Hsiao, Liang-Lin. China's Foreign Trade Statistics, 1864-1949. Cambridge: Harvard University Press, 1974.

Japan-Manchoukuo Year Book, 1934.

Jen-min Shou-t'se [People's Handbook]. People's Republic of China. Tientsin and Shanghai: Ta Kung Pao She, 1949-1966.

Kuang-hsu san-shih-san nien yu-ch'uan-pu ti-i-tz'u, chi ti-erh-tz'u t'ung-chi-piao. (Statistical Tables of the Ministry of Posts and Communications, first issue, for 1907 and second issue, for 1908). Republic of China. Peking: 1910, 6 ts'e: 1911, 8 ts'e.

Kuo-min cheng-fu nien-chien [Yearbook of National Government], Republic of China, Executive Yuan. n.p., Wen-yi-nan Printing, 1943, 1944, 1946.

Li Choh-ming. The Statistical System of Communist China. Berkeley: University of California Press, 1962.

Statistics of Government Railways in China, 1915-1921: retitled Statistics of Railways, 1922-1927; and Statistics of Chinese National Railways, 1928-. Republic of China, Ministry of Railways, Peking and Nanking.

T'ieh-lu nien-chien [The railway yearbook]. Republic of China, Shanghai: Commercial Press, 1933.

Tsui-chin san-shih-shi nien lai chung-kuo tung hsiang kou-an tui-wai mo-yi tung-chi: chung-pu [Statistics of China's foreign trade by ports, 1900-1933: 1. Central ports] Republic of China, Ministry of Industry. Shanghai: Commercial Press, 1935.

T'ung-chi kung-tso [Statistical Work], no. 11, 1957.

Yen Chung-p'ing, et al. Chung-kuo chin-tai ching-chi-shih t'ung-chi tzu-liao hsuan-chi [Selection of statistical materials on Chinese modern economic history, 1760-1949]. Peking: Ko Hsueh chu-pan-she, 1955.

Yang, T. L. and Hau, H. B., et al. Liu-shih-wu nien lai chung-kuo kuo-chi mo-yi tung-chi [Statistics of China's foreign trade during the last sixty-five years]. National Research Institute of Social Sciences, Academia Sinica, Monograph no. 4, 1931.

Yin, Helen and Yin, Yi-chang. Economic Statistics of Mainland China 1949-1957. Cambridge: Harvard University Press, 1960.

Yuan Wen-chang. Tung-pei tieh-lu wen-ti [Railway problems in the Northeast]. Shanghai: Chunghua, 1932.

D. British Foreign Office's Files

(This list includes only some of the files actually consulted)

General Correspondence, 1815-

F.O.	17	470	471	779	1757-1764
F.O.	371	8033	15445	15446	16206
		16207	18053	18130	19250
		19249	19252	20972	23433

Embassy and Consular Archives, 1834-

F.O.	228	2127	2134	2140	2141
		2267-3386	2292	2298	2386-2391
		2470-2474	2492-2495	2521-2530	2590
F.O.	233	47	78	79	237-239
		244	251	254	256
		259	117/15	119/6	119/9
		120/106	122/14	122/29	122/30
		123/4	123/5-9	123/13	123/23
		125/31	126/6	126/62	127/8
		127/30	128/21-22	128/24	128/40
		128/45	128/47	128/55	129/2-1
		129/9	129/13	129/22	219/37
		129/43	130/14	130/20	130/31
		130/33-34	130/45	130/50	131/3

Embassy and Consular Archives, 1834-

F.O.	233	131/11	131/21	131/32	131/37
		131/45	131/49	131/52	132/6
		132/8	132/10	132/14	132/40
		132/54	132/55	132/60	132/71
		133/6	133/7	133/15	133/32
		133/33	134/40	134/77	

Confidential Print, c. 1848-

F.O. 405 180-181 188-189 197-198 202-203

E. Maps, Atlases, and Political Areas

Chang Ch'i-yun. Chung-hua-min-kuo ti t'u chi [Atlas of the Republic of China]. Taipei: Institute of National Defence, 1959.

China (PRC). Ministry of Interior Affairs. Chung-hua jen-min kung-ho kuo hsing-cheng ch'u-hua shou-ts'e. Peking: Ti-tu chu-pan she, 1965.

________. Ministry of Interior Affairs. Chung-hua jen-min kung-ho kuo hsing-cheng ch'u-hua hien-ts'e. Peking: Fa-lu ch'u-pan she, 1958.

________. Chung-hua jen-min kung-ho-kuo ti-tu [Map of the People's Republic of China]. 1:6,000,000, Peking: 1965 edition, 7th printing, January 1973.

________. Zhonghua Remin Gongheguo Ditu [Map of the People's Republic of China]. 1:6,000,000, Peking: June 1974, first edition, first printing.

________. Chung-hua jen-min kung-ho-kuo fun sh'eng ti-tu [Provincial atlas of the People's Republic of China]. Peking, Ti-tu Chu-p'an she, October 1974.

________. Chung-kuo ti-shing, 1:4,000,000 (Wall Map) [Relief map of China]. Peking: Ti-tu chu-pan she, March 1974.

China (PRC). Ministry of the Interior. Chung-hua min-kuo hsing-cheng Ch'u-yu chien-piao. Shanghai: Commercial Press, 1947.

________. Ministry of the Interior. Chung-kuo chih hsing-cheng tu-ch'a ch'u. Shanghai: Ta Chung Publishing Co., 1948.

Chou Shih-t'ang. Erh-shih shih-chi chung wai ta ti-t'u [Map of 20th Century China and the World]. n.p., 1966.

Chung-hua tsui hsin ti-tu [Newest map of China]. n.p., n.d.

Fang, Ch'iu-wei. Chung-hua jen-min kung-ho-kuo fen-sheng ti-t'u chi [Atlas of rovinces of the People's Republic of China]. Peking: Ti-t'u ch'u-pan-she, 1974.

Fu Chueh-chin, Chung-hua min-kuo hsing-cheng ch'u-yu t'u [Administrative map of the Republic of China]. n.p. 1947.

Min Pe-ch'ien, Hsin-pien chung-kuo chuan-tu [New wall map of China]. n.p., 1939.

Herrmann, Albert. Historical and Commercial Atlas of China. Cambridge: Harvard University Press, 1935.

________. An Historical Atlas of China. Partly Revised and edited by Norton Ginsburg. Chicago: Aldine Publishing Company, 1966.

Hsieh, I-yuan. "Chung-hua jen-min kung-he kuo hsing-cheng ch'u-yu ti hua-fen." [Delineation of administrative regions of the PRC]. Acta Geographica Sinica 24 (February 1958): 84-97.

Hsieh, Pin. Chung-kuo sang ti shih [History of China's losses of territories]. Shanghai: China Books, 1936.

Teng, Yen-lin. Chung-kuo pien-chiang t'u chi-lu [A Bibliography of maps, atlases, and books of China's borderlands] Shanghai: Commercial Press, 1958.

Ting Wen-chiang, ed. Chung-kua-min-kuo hsin ti-t'u [New tlas of the Republic of China]. Shanghai: Shen Pao Press, 1934.

U.S.A. Central Intelligence Agency. China: provisional atlas of Communist administrative units. Washington, D.C.: Central Intelligence Agency, 1959.

________. Communist China Map Folio. October 1967.

________. Communist China Administrative Atlas. March 1969.

________. Atlas, People's Republic of China, Washington, D.C.: Central Intelligence Agency, November 1971.

U.S.A. Military Intelligence Division, U.S. Army, Railroads of China, Vols. 1-7. Washington: Strategic Engineering Study, no. 145, November 1944.

Yang, Yu-liu. Chung-kuo li-tai ti-fang hsing-cheng ch'u-hua [Local dministrative ivision in China]. Taipei: China Publishing Company, 1957.

Yeh, Hsiang-chih. Kung-fei ch'ieh chu hsia ti chung-kuo ta lu fen sheng ti t'u [Provincial atlas of Communist China]. Taipei: Institute of National Defence, 1966.

II. SECONDARY SOURCES

A. Books

Abler, Ronald, Adams, John S. and Gould, Peter. Spatial Organization. Englewood Cliffs: Prentice-Hall, 1971.

Aird, John S. The Size, Composition, and Growth of the Population of Mainland China. U.S. Bureau of the Census, International Population Statistics Reports, series P-90, no. 15, Washington D.C., 1961.

________. Estimates and Projections of the Population of Mainland China: 1953-1986. U.S. Bureau of the Census, Washington, D.C., no. 17, 1968.

Appleton, J. H. A Morphological Approach to the Geography of Transport. Hull: Universtiy of Hull, 1965.

Beresford, Charles. The Break-Up of China. London: Harper and Brothers, 1899.

Bland, J. O. P. Li Hung-chang. New York: Henry Holt and Co., 1917. (A biography of Li with a shade of negation).

Buchanan, Colin. Traffic in Towns. London: Britain Ministry of Transport, 1963.

Buchanan, Keith McPherson. The Transformation of the Chinese Earth. London: G. Bell & Sons, 1970.

Buckholts, Paul. Political Geography. New York: Ronald Press, 1966.

Carlson, Lucile. Geography and World Politics. Englewood Cliffs: Prentice-Hall, 1958.

Chang, Wai-ya. K'ung-fei ch'ai cheng ch'in-yung chi yen-ch'iu [A Study of Communist China's finance and monetary system]. Taipei: Yangming Shan-chuang, 1957.

________. Chung-k'ung wu-nien chi-hua po-shih [An analysis of Communist China's First Five-Year Plan]. Hong Kong: Freedom Press, 1955.

Chiang Kai-shek. China's Destiny. New York: Macmillan Company, 1947.

Chinese Academy of Science. Wo-kuo ching-chi-hsueh-chieh kuan-yu she-hui tzu-yi chi-t'u hsia hsiang-p'in, chia-chi ho chia-ke wen-ti lun-wen hsuan-chi [Selected essays on commodities, value, pricing system under the Socialist system by China's economists]. Peking: Scientific Press, 1958.

Chow, C. T. "China's Internal Railway Problems: The Case of the Railways' First Century, 1866-1966." Ph.D. Dissertation, Michigan State University, 1972.

Clyde, Paul Hibbert. International Rivalries in Manchuria, 1689-1922. Columbus: The Ohio State University Press, 1926.

Cohen, Saul B. Jerusalem: Bridging the Four Walls. New York: Herzl Press, 1977.

Colquhoun, Archibald R. China in Transformation. London: Harper, 1898.

Conolly, Violet. Siberia Today and Tomorrow. London: Collins Press, 1975.

Cooley, Charles H. The Theory of Transportation. Baltimore: American Economic Association Publication, 1894.

Dawson, Owen L. Communist China's Agriculture. New York: Frederick A. Praeger, 1970.

Dennett, T. Americans in East Asia. New York: Barnes and Noble, 1922.

Deutsch, Karl W. Nationalism and Social Communication: Inquiry into the Foundations of Nationality. Cambridge: M.I.T. Press, 1966.

Donnithorne, Audrey Gladys. China's Economic System. New York: Frederick A. Praeger, 1967.

Eliot Hurst, Michael E. Transportation Geography, Comments and Readings. New York: McGraw-Hill Book Company, Inc., 1974.

Etzioni, Amitai, ed. Political Unification: A Comparative Study of Leaders and Forces. New York: Holt, Rinehart, and Winston, 1965.

Fang Hao. Chung-kuo ch'in-tai wai-chiao shih. [A history of China's modern diplomacy]. Taipei: Chinese Culture Publishing Enterprise Committee, 1955.

Fishlow, Albert. American Railroads and the Transformation of the Ante-Bellum Economy. Cambridge: Harvard University Press, 1965.

Fogel, Robert W. Railroads and American Economic Growth. Baltimore: Johns Hopkins Press, 1964.

Feuerwerker, Albert. China's Early Industrialization. Cambridge: Harvard University Press, 1958.

________. China's Early Modernization, Sheng Hsuen-huai (1844-1916) and Mandarin Enterprise. Harvard East Asian Series, no. 1. Cambridge: Harvard University Press, 1958.

Fuchs, V. R. Changes in the Location of Manufacturing in the United States Since 1929. New Haven: Yale University Press, 1962.

Garrison, William L., and Marble, Dwayne F. The Structure of Transportation Networks. U.S. Army Transportation Command, Technical Report 62-11, 1962.

Garrison, William L. et al. Studies of Highway Development and Geographic Change. Seattle: University of Washington Press, 1959.

Ginsburg, Norton. The Pattern of Asia. Englewood Cliffs: Prentice-Hall, 1958.

Gleditsch, Nils Peter. The Structure of the International Airline Network. Oslo: mimeographed, 1968.

Goodrich, Carter. Government Promotion of American Canals and Railroads, 1800-1890. New York: Columbia University Press, 1960.

Gould, Peter R. The Development of the Transportation Pattern in Ghana. Studies in Geography, no. 5. Evanston: Northwestern University Press, 1960.

Grodinsky, Julius. Transcontinental Railway Strategy, 1869-1893: A Study of Businessmen. Philadelphia: University of Pennsylvania Press, 1962.

Haefele, Edwin T., ed. Transport Planning and National Goals. Washington, D.C.: Brookings Institution Press, 1960.

Haggett, Peter. Locational Analysis in Human Geography. London: Edward Arnold Publishers Ltd., 1965.

Haggett, Peter and Chorley, Richard J. Network Analysis in Geography. New York: St. Martin's Press, 1969.

Hay, Alan. Transport for the Space Economy: A Geographical Study. Seattle: University of Washington Press, 1973.

Ho Ping-ti. Studies on the Population of China 1368-1953. Cambridge: Harvard University Press, 1959.

Hoover, Edgar M. The Location of Economic Activity. New York: McGraw-Hill Book Company, Inc., 1948.

Horton, Frank, ed. Geographic Studies of Urban Transportation and Network Analysis. Studies in Geography no. 16. Evanston: Northwestern University Press, 1968.

Hoyle, B. S. Transport and Development. New York: Barnes and Noble Press, 1973.

Hsu, Shushi. An Introduction to Sino-Foreign Relations. Shanghai: Kelly and Walsh, 1941.

Hunt, Michael H. Frontier Defense and the Open Door. New Haven: Yale University Press, 1973.

Hunter, Holland. Soviet Transport Experience: Its Lessons for Other Countries. Washington, D.C.: Brookings Institution, 1968.

________. Chinese and Soviet Transport for Agriculture. University of Pittsburgh, UCIS Occasional Papers, 1974.

Iriye, Akira. Across the Pacific. New York: Harcourt, Brace and World, 1967.

________. After Imperialism: The Search for a New Order in the Far East 1921-1931. New York: Atheneum, 1969.

Isard, Walter. Location and Space-Economy: A General Theory Relating to Industrial Location, Market Areas, Land Use, Trade and Urban Structure. Cambridge: M.I.T. Press, 1956.

Jones, F. C. Manchuria Since 1931. London: Oxford University Press, 1949.

Kansky, Karel J. Structure of Transportation Networks: Relationship Between Network Geometry and Regional Characteristics. Research Paper no. 84. Chicago: University of Chicago, Department of Geography, 1963.

Kinnosuke, Adachi. Manchuria. New York: Robert M. McBride, 1925.

Lansing, John B. Transportation and Economic Policy. New York: The Free Press, 1966.

Lattimore, Owen. Inner Asian Frontiers of China. New York: American Geographical Society, 1940.

LeFevour, Edward. Western Enterprise in Late Ch'ing China. A Selective Survey of Jardine, Matheson and Company's Operations, 1842-1895. Cambridge: Harvard University Press, 1968.

Leinbach, Thomas R. "Transportation and the Development of Malaya." Annals of the Association of American Geographers 65 (June 1975): 270-282.

Leipmann, Kate K. The Journey to Work. London: n.p., 1945.

Lewis, R. S. Eighty Years of Enterprise, 1869-1949, Being the Ultimate Story of the Waterside Works of Ransomes and Papier Ltd. of Ipswich. England, Ipswich, n.p.

Li Kuo-chi. Chang chih-tung ti wai-chiao cheng-ch'e [Chang chih-tung's foreign policy]. Taipei: Academia Sinica, Institute of Modern History, 1970.

Lindbeck, J. M. H., ed. China, Management of a Revolutionary Society. Seattle: University of Washington Press, 1971.

Liu, Alan P. L. Communications and National Integration in Communist China. New York: John Wiley and Sons, 1964.

Liu, Ta-chung and Yeh, Kung-chia. The Economy of the Chinese Mainland: National Income and Economic Development, 1933-1959. Santa-Monica: Rand Corporation, April 1963.

Lösch, August. The Economics of Location. New Haven: Yale University Press, 1959.

MacMurray, John V. A., ed. Treaties and Agreements With and Concerning China, 1894-1919. London: Oxford University Press, 1921.

Mahan, A. T. The Problem of Asia. Boston: Little, Brown and Company, 1900.

Mazlish, Bruce, ed. The Railroad and the Space Program: An Exploration in Historical Analogy. Cambridge: The M.I.T. Press, 1965.

Moodie, A. E. Geography Behind Politics. London: Hutchinson Press, 1949.

Morrill, Richard L. Migration and the Growth of Urban Settlement. Lund Studies in Geography, no. 26. Lund: Royal University of Lund, 1965.

Moseley, George V. H., III. The Consolidation of the South China Frontier. Berkeley: University of California Press, 1973.

Mossman, Frank H. and Newton, Morton. Principles of Transportation. New York: Ronald Press, 1957.

Nelson, James C. Railroad Transportation and Public Policy. Washington, D.C.: Brookings Institution, 1959.

Ni, E. Distribution of the Urban and Rural Population of Mainland China: 1953 and 1958. U.S. Bureau of the Census, International Population Statistics Reports, Series P-95, no. 56, October 1960.

Nie, Norman H. et al. Statistical Package for the Social Sciences. New York: McGraw-Hill, 1970, pp. 174-195.

O'Connor, Anthony M. Railways and Development in Uganda, A Study in Economic Geography. Nairobi: Oxford University Press, 1965.

O'Dell, A. C. Railways and Geography. London: Hutchinson University Press, 1956.

Oksenberg, Michel, ed. China's Development Experience. New York: Frederick A. Praeger, 1973.

Orleans, Leo A. Every Fifth Child: the Population of China. Stanford: Stanford University Press, 1972.

O'Sullivan, Patrick. Transport Networks and the Irish Economy. London: London School of Economics and Political Science Press, 1969.

Owen, Wilfred. Distance and Development: Transport and Communications in India. Washington D.C.: Brookings Institution, 1968.

________. Strategy for Mobility. Washington, D.C.: Brookings Institution Press, 1964.

Pao, Ming-chien J. The Open Door Doctrine. New York: Macmillan Company, 1923.

Perkins, D. H. Agricultural Development in China, 1368-1968. Chicago: Aldine Press, 1969.

Potts, R. B. and Oliver, R. M. Flows in Transportation Networks. New York: Academic Press, 1972.

Pye, Lucian. Communication and Political Development. Princeton: Princeton University Press, 1963.

Rimmer, Peter J. Transport in Thailand. Department of Human Geography. Canberra: Australian National University Press, 1971.

Ripley, William Z., ed. Railway Problems. Boston: Ginn and Co., 1907.

Rockhill, William W. Treaties and Conventions With or Concerning China and Korea, 1894-1904. Washington, D.C.: n.p., 1904.

Rostow, Walter W. The Stages of Economic Growth. Cambridge: Cambridge University Press, 1964.

Ruppenthal, Karl M., ed. Challenge to Transportation. Stanford: Stanford University Press, 1961.

Schram, Stuart. The Political Thought of Mao Tse-tung. New York: Frederick A. Praeger, 1963.

________, ed. Mao Tse-tung Unrehearsed. Harmondsworth: Penguin Books, 1974.

de Seversky, Alexander P. Air Power: Key to Survival. New York: Simon and Schuster, 1950.

Shan Chaun Leng and Palmer, Norman D. Sun Yat-sen and Communism. London: Thames and Hudson, 1961.

Shen, T. H. Agricultural Resources of China. New York: Cornell University Press. 1951. (Also published in Chinese by China Culture Publishing Enterprise Committee, Taipei, 1952).

Shinobu Seizaburo, Nihon Gaiko-shi, 1853-1972. Tokyo: Mainich Shinbun-sha, 1974.

Soja, Edward W. The Political Organization of Space. Washington, D.C.: Association of American Geographers, 1971.

Taaffe, Edward J. and Gauthier, Howard L. Jr. Geography of Transportation. Englewood Cliffs: Prentice-Hall, 1973.

Tan, Chester. Chinese Political Thought in the Twentieth Century. Garden City: Doubleday, 1971.

T'ang Erh-ho. Pei-man kai-lun [Outline of the economy of North Manchuria]. Translated from Japanese. Shanghai: Commercial Press, 1937.

Thomas, Frank H. The Denver and Rio Grande Western Railroad: A Geographic Analysis. Studies in Geography, no. 4. Evanston: Northwestern University Press, 1960.

Tomimas, Shutaro. The Open Door Policy and Territorial Integrity of China. New York: Red House Press, 1918.

Tsou, Tang. America's Failure in China, 1941-50. 2 vols. Chicago: University of Chicago Press, 1967.

Tuan Yi-fu. China. London: Longmans, 1970.

U.S. Congress. Joint Economic Committee. An Economic Profile of Mainland China. New York: Frederick A. Praeger, 1968.

________. Joint Economic Committee. People's Republic of China: An Economic Assessment. Washington, D.C: U.S. Government Printing Office, 1972.

U.S. Social Science Research Council. Committee on the Economics of China. Provincial Agricultural Statistics for Communist China. Ithaca, N.Y.: the Council, 1969.

Wellington, Arthur Mellen. The Economic Theory of the Location of Railways. New York: J. Wiley & Sons, 1908.

Whitney, Joseph B. R. China: Area, Administration, and Nation Building. Department of Geography Research Papers, no. 123. Chicago: University of Chicago Press, 1970.

Whittlesey, Derwent. The Earth and the State. New York: Henry Holt, 1944.

Wolfe, Roy I. Transportation and Politics. New Jersey: D. Van Nostrand Co. Inc., 1963.

Wu Yuan-li. The Economy of Communist China. New York: Frederick A. Praeger, 1965.

Yeates, Maurice. An Introduction to Quantitative Analysis in Human Geography. New York: McGraw-Hill, 1974.

B. Articles

Ackerman, William V. "Development Strategy for Cuyo, Argentina." Annals of the Association of American Geographers 65 (March 1975): 36-47.

Baker, J. E. "Transportation in China." Annals of the American Academy of Political and Social Science 152 (November 1930): 168.

Berry, Brian J. L. "An Inductive Approach to the Regionalization of Economic Development." In Essays in Geography and Economic Development, pp. 78-107. Edited by Norton Ginsburg. Chicago: University of Chicago Press, 1960.

________. "Cities and Systems Within Systems of Cities." In Regional Development and Planning, pp. 116-137. Edited by W. Alonso and J. Friedmann. Cambridge: M.I.T. Press, 1964.

________. "Recent Studies Concerning the Role of Transportation in the Space Economy." Annals of the Association of American Geographers 49 (December 1959): 328-341.

Bisson, T. A. "Railway Rivalries in Manchuria Between China and Japan." Foreign Policy Reports 8 (13 April 1932): 29-42.

Briggs, J. A. "The Development of the East African Railway Network - An Application of a Sequential Development Model." Swansea Geographer 12 (1974): 47-49.

Brookfield, H. C. "On One Geography and A Third World." Transactions 58 (March 1973): 1-20.

Chan Tieh-chang. "Tabulation Forms for Transportation Planning." Chi-hua ching-chi [Planned Economy]. 3 March 1957, pp. 29-33.

Chang Pang-ying. "Transport Undertaken by People's Communes." Red Flag, no. 17, (September 1959), p. 31.

Chang Shih. "A Major Problem in Promoting a Rational System of Coal Transport." Planned Economy, no. 10, (October 1957), p. 24.

Chang Wu-tung and Yang Kuan-hsiung. "Function of Transport in the Development of Agricultural Production." People's Daily, 6 June 1964, p. 5.

Ch'iu Hsien-hung. "Improve Field Transport and Save Labor." People's Daily, 8 December 1960, p. 4.

Cootner, P. "The Role of the Railroads in the U.S. Economic Growth." Journal of Economic History 23 (December 1963): 477-521.

Dickinson, Robert E. "The Geography of Commuting: The Netherlands and Belgium." Geographical Review 47 (October 1957): 521-538.

________. "The Geography of Commuting in West Germany." Annals of the Association of American Geographers 49 (December 1959): 443-456.

Eliot Hurst, Michael E. "Transportation and the Societal Framework." Economic Geography 49 (April 1973): 163-180.

Fisher, C. A. "The Railway Geography of British Malaya." The Scottish Geographic Magazine, Vol. 64, no. 3 (1948), pp. 123-136.

Forer, P. C. "Relative Space and Regional Imbalance: Domestic Airlines in New Zealand's Geometrodynamics." Proceedings of the International Geographical Union Regional Conference, New Zealand, 1974, pp. 53-62.

Friedmann, J. "An Approach to Policies Planning for Spatial Development." School of Architecture and Urban Planning. Los Angeles: University of California at Los Angeles, July, 1974.

Fuchs, V. R. "Statistical Explanations of the Relative Shift of Manufacturing Among Regions of the United States." Papers and Proceedings of the Regional Science Association 8, 1962, pp. 105-126.

Fulton, Maurice and Hoch, L. Clinton. "Transportation Factors Affecting Location Decisions." Economic Geography 35 (January 1959): 51-59.

Gauthier, Howard L. "Geography, Transportation and Regional Development." In Transport and Development, pp. 19-31. Edited by B. S. Hoyle. New York: Barnes and Noble Press, 1973.

Ginsburg, Norton. "Ch'ang-ch'un." Economic Geography 23 (October 1947): 290-307.

________. "China's Changing Political Geography." Geographical Review 42 (January 1952): 102-117.

________. "The Political Dimension: Regionalism and Extra-regional Relations in S.E. Asia." Focus 23 (December 1972): 1-8.

Gottmann, Jean. "Geography and International Relations." World Politics 3 (March 1951): 153-173.

________. "The Political Partitioning of the World: An Attempt of Analysis." World Politics 4 (July 1952): 512-519.

Hartshorne, Richard. "The Functional Approach in Political Geography." Annals of the Association of American Geographers 40 (June 1950): 95-130.

________. "Political Geography." In American Geography: Inventory and Prospect, pp. 167-225. Edited by Preston E. James and Clarence F. Jones. Syracuse: Syracuse University Press, 1954.

Hebert, Budd and Murphy, E. "Evolution of an Accessibility Surface: The Case of the U.S. Air Transport Network." Proceedings of the Association of American Geographers, vol. 3, (1971), pp. 75-79.

Herman, Theodore. "Group Values Toward the National Space: the Case of China." Geographical Review 49 (April 1959): 164-182.

Hilling, David. "Politics and Transportation: the Problems of West Africa's Landlocked States." In Essays in Political Geography, pp. 253-269. Edited by C. A. Fisher. London: Methuen Press, 1968.

Huenemann, Ralph W. and Ludlow, Nicholas H. "China's Railroads." The China Business Review (March-April 1977), pp. 26-43.

Hunter, Holland. "Transport in Soviet and Chinese Development." Economic Development and Cultural Change 14 (1965): 71-72.

Jefferson, M. "The Civilizing Rails." Economic Geography 4 (October 1928): 217-231.

Jones, Stephen B. "A Unified Field Theory of Political Geography." Annals of the Association of American Geographers 44 (June 1954): 111-123.

Kan, T. "Transportation in China." Studies on Chinese Communism 60 (February 1972).

Kolars, John and Malin, Henry J. "Population and Accessibility: An Analysis of Turkish Roads." Geographical Review 60 (April 1970): 229-246.

Kristof, Ladis K. "The Origins and Evolution of Geopolitics." Conflict Resolution. 4 (March 1960): 15-51.

________. "The State-Idea, and National Idea and the Image of the Fatherland." Orbis 11 (September 1967): 238-255.

Lei, T'ing and Chu, Hsing-min. "Tuan-t'u yun-shu ti t'e-tien chi ch'i tsai ch'eng-hsiang wu-tzu chiao-liu-chung ti tso-yung." [The characteristics of short-distance transportation and its function in the commodity flow between rural and urban areas]. Ching-chi Yen-chiu. no. 6, 1963, Peking, p. 33.

Leinbach, Thomas R. "Transportation and the Development of Malaya." Annals of the Association of American Geographers 65 (June 1975): 270-282.

Li Po-chih. "Kao-hao tuan-t'u yun-shu" [Improving short-distance transportation]. Chi-hua yu ung-chi, [Planning & statistics], Peking, 7 (April 1959): 24.

Lonsdale, Richard E. "Two North Carolina Commuting Patterns." Economic Geography 42 (April 1966): 114-138.

MacKinder, H. J. "The Geographical Pivot of History." Geographical Journal. 23 (1904): 421-444.

Moseley, George. "The Frontier Regions in China's Recent International Politics." In Modern China's Search for a Political Form, pp. 299-329. Edited by J. Gray. London: Oxford University Press, 1969.

Nystuen, John D. and Dacey, Michael F. "A Graph Teory Interpretation of Nodal Regions." In Spatial Analysis, pp. 407-417. Edited by Brian J. L. Berry and Dwayne F. Marble. Englewood Cliffs: Prentice-Hall, 1968.

Pavlov, K. "China's Railroads." Bulletin 4 (September 1963): 12-28.

Penrose, E. G. "The Place of Transport in Economic and Political Geography." UN Transport and Communications Review 5 (April-June 1952): 1-8.

Rosenbaum, Arthur. "Railway Enterprise and Economic Development, the Case of the Imperial Railways of North China, 1900-1911." Modern China 2 (1976): 227-272.

Roxby, P. M. "China as an Entity: The Comparison with Europe." Geography 19 (March 1934): 1-20.

Shimbel, Alfonso. "Structural Parameters of Communication Networks." Bulletin of Mathematical Biophysics 15 (1953): 501-507.

Soja, Edward W. "Communications and Territorial Integration in East Africa." The East Lakes Geographer 4 (December 1968): 39-57.

Stanley, W. R. "Transport Expansion in Liberia." Geographical Review 60 (October 1970): 529-547.

Tsou Tang. "Western Concepts and China's Historical Experience." World Politics 21 (July 1969): 655-691.

Taaffe, Edward J. "The Transportation Network and the Changing American Landscape." In Problems and Trends in American Geography, pp. 15-25. Edited by Saul B. Cohen. New York: Basic Books, 1967.

Taaffe, Edward J., Morrill, Richard L. and Gould, Peter R. "Transportation Expansion in Underdeveloped Countries: A Comparative Analysis." Geographical Review 53 (October 1963): 503-529.

Teiwes, Frederick C. "Provincial Politics in China: Themes and Variations." In China Management of a Revolutionary Society, pp. 117-185. Edited by John M. H. Lindbeck. Seattle: University of Washington Press, 1971.

Ullman, Edward L. "The Role of Transportation and the Bases of Interaction." In Man's Role in Changing the Face of the Earth, pp. 862-895. Edited by William L. Thomas. Chicago: University of Chicago Press, 1956.

________. "Transportation Geography." In American Geography: Inventory and Prospect, pp. 310-332. Edited by Preston E. James and Clarence F. Jones. Association of American Geographers, 1954.

Wallace, W. H. "Railroad Traffic Densities and Patterns." Annals of the Association of American Geographers 48 (December 1958): 352-374.

Ward, M. "Progress in Transport Geography." In Trends in Geography, pp. 164-173. Edited by R. U. Cooke and J. H. Johnson. Oxford: Pergamon Press, 1969.

THE UNIVERSITY OF CHICAGO
DEPARTMENT OF GEOGRAPHY
RESEARCH PAPERS (Lithographed, 6×9 inches)

Available from Department of Geography, The University of Chicago, 5828 S. University Avenue, Chicago, Illinois 60637, U.S.A. Price: $8.00 each; by series subscription, $6.00 each.

LIST OF TITLES IN PRINT

48. BOXER, BARUCH. *Israeli Shipping and Foreign Trade.* 1957. 162 p.
56. MURPHY, FRANCIS C. *Regulating Flood-Plain Development.* 1958. 204 p.
62. GINSBURG, NORTON, editor. *Essays on Geography and Economic Development.* 1960. 173 p.
69. CHURCH, MARTHA. *Spatial Organization of Electric Power Territories in Massachusetts.* 1960. 187 p.
71. GILBERT, EDMUND WILLIAM *The University Town in England and West Germany.* 1961. 79 p.
72. BOXER, BARUCH. *Ocean Shipping in the Evolution of Hong Kong.* 1961. 108 p.
74. TROTTER, JOHN E. *State Park System in Illinois.* 1962. 152 p.
79. HUDSON, JAMES W. *Irrigation Water Use in the Utah Valley, Utah.* 1962. 249 p.
91. HILL, A. DAVID. *The Changing Landscape of a Mexican Municipio, Villa Las Rosas, Chiapas.* 1964. 121 p.
92. SIMMONS, JAMES W. *The Changing Pattern of Retail Location.* 1964. 200 p.
97. BOWDEN, LEONARD W. *Diffusion of the Decision to Irrigate: Simulation of the Spread of a New Resource Management Practice in the Colorado Northern High Plains.* 1965. 146 p.
98. KATES, ROBERT W. *Industrial Flood Losses: Damage Estimation in the Lehigh Valley.* 1965. 76 p.
101. RAY, D. MICHAEL. *Market Potential and Economic Shadow: A Quantitative Analysis of Industrial Location in Southern Ontario.* 1965. 164 p.
102. AHMAD, QAZI. *Indian Cities: Characteristics and Correlates.* 1965. 184 p.
103. BARNUM, H. GARDINER. *Market Centers and Hinterlands in Baden-Württemberg.* 1966. 172 p.
104. SIMMONS, JAMES W. *Toronto's Changing Retail Complex.* 1966. 126 p.
105. SEWELL, W. R. DERRICK, *et al. Human Dimensions of Weather Modification.* 1966. 423 p.
106. SAARINEN, THOMAS FREDERICK. *Perception of the Drought Hazard on the Great Plains.* 1966. 183 p.
107. SOLZMAN, DAVID M. *Waterway Industrial Sites: A Chicago Case Study.* 1967. 138 p.
108. KASPERSON, ROGER E. *The Dodecanese: Diversity and Unity in Island Politics.* 1967. 184 p.
109. LOWENTHAL, DAVID, editor, *Environmental Perception and Behavior.* 1967. 88 p.
110. REED, WALLACE E., *Areal Interaction in India: Commodity Flows of the Bengal-Bihar Industrial Area.* 1967. 209 p.
112. BOURNE, LARRY S. *Private Redevelopment of the Central City, Spatial Processes of Structural Change in the City of Toronto.* 1967. 199 p.
113. BRUSH, JOHN E., and GAUTHIER, HOWARD L., JR., *Service Centers and Consumer Trips: Studies on the Philadelphia Metropolitan Fringe.* 1968. 182 p.
114. CLARKSON, JAMES D., *The Cultural Ecology of a Chinese Village: Cameron Highlands, Malaysia.* 1968. 174 p.
115. BURTON, IAN, KATES, ROBERT W., and SNEAD, RODMAN E. *The Human Ecology of Coastal Flood Hazard in Megalopolis.* 1968. 196 p.
117. WONG, SHUE TUCK, *Perception of Choice and Factors Affecting Industrial Water Supply Decisions in Northeastern Illinois.* 1968. 93 p.
118. JOHNSON, DOUGLAS L.. *The Nature of Nomadism: A Comparative Study of Pastoral Migrations in Southwestern Asia and Northern Africa.* 1969. 200 p.
119. DIENES, LESLIE. *Locational Factors and Locational Developments in the Soviet Chemical Industry.* 1969. 262 p.
120. MIHELIČ, DUŠAN. *The Political Element in the Port Geography of Trieste.* 1969. 104 p.
121. BAUMANN, DUANE D. *The Recreational Use of Domestic Water Supply Reservoirs: Perception and Choice.* 1969. 125 p.
122. LIND, AULIS O. *Coastal Landforms of Cat Island, Bahamas: A Study of Holocene Accretionary Topography and Sea-Level Change.* 1969. 156 p.
123. WHITNEY, JOSEPH B. R. *China: Area, Administration and Nation Building.* 1970. 198 p.
124. EARICKSON, ROBERT. *The Spatial Behavior of Hospital Patients: A Behavioral Approach to Spatial Interaction in Metropolitan Chicago.* 1970. 138 p.
125. DAY, JOHN CHADWICK. *Managing the Lower Rio Grande: An Experience in International River Development.* 1970. 274 p.

126. MacIVER, IAN. *Urban Water Supply Alternatives: Perception and Choice in the Grand Basin Ontario.* 1970. 178 p.

127. GOHEEN, PETER G. *Victorian Toronto, 1850 to 1900: Pattern and Process of Growth.* 1970. 278 p.

128. GOOD, CHARLES M. *Rural Markets and Trade in East Africa.* 1970. 252 p.

129. MEYER, DAVID R. *Spatial Variation of Black Urban Households.* 1970. 127 p.

130. GLADFELTER, BRUCE G. *Meseta and Campiña Landforms in Central Spain: A Geomorphology of the Alto Henares Basin.* 1971. 204 p.

131. NEILS, ELAINE M. *Reservation to City: Indian Migration and Federal Relocation.* 1971. 198 p.

132. MOLINE, NORMAN T. *Mobility and the Small Town, 1900–1930.* 1971. 169 p.

133. SCHWIND, PAUL J. *Migration and Regional Development in the United States.* 1971. 170 p.

134. PYLE, GERALD F. *Heart Disease, Cancer and Stroke in Chicago: A Geographical Analysis with Facilities, Plans for 1980.* 1971. 292 p.

135. JOHNSON, JAMES F. *Renovated Waste Water: An Alternative Source of Municipal Water Supply in the United States.* 1971. 155 p.

136. BUTZER, KARL W. *Recent History of an Ethiopian Delta: The Omo River and the Level of Lake Rudolf.* 1971. 184 p.

139. McMANIS, DOUGLAS R. *European Impressions of the New England Coast, 1497–1620.* 1972. 147 p.

140. COHEN, YEHOSHUA S. *Diffusion of an Innovation in an Urban System: The Spread of Planned Regional Shopping Centers in the United States, 1949–1968,* 1972. 136 p.

141. MITCHELL, NORA. *The Indian Hill-Station: Kodaikanal.* 1972. 199 p.

142. PLATT, RUTHERFORD H. *The Open Space Decision Process: Spatial Allocation of Costs and Benefits.* 1972. 189 p.

143. GOLANT, STEPHEN M. *The Residential Location and Spatial Behavior of the Elderly: A Canadian Example.* 1972 226 p.

144. PANNELL, CLIFTON W. *T'ai-chung, T'ai-wan: Structure and Function.* 1973. 200 p.

145. LANKFORD, PHILIP M. *Regional Incomes in the United States, 1929–1967: Level, Distribution, Stability, and Growth.* 1972. 137 p.

146. FREEMAN, DONALD B. *International Trade, Migration, and Capital Flows: A Quantitative Analysis of Spatial Economic Interaction.* 1973. 201 p.

147. MYERS, SARAH K. *Language Shift Among Migrants to Lima, Peru.* 1973. 203 p.

148. JOHNSON, DOUGLAS L. *Jabal al-Akhdar, Cyrenaica: An Historical Geography of Settlement and Livelihood.* 1973. 240 p.

149. YEUNG, YUE-MAN. *National Development Policy and Urban Transformation in Singapore: A Study of Public Housing and the Marketing System.* 1973. 204 p.

150. HALL, FRED L. *Location Criteria for High Schools: Student Transportation and Racial Integration.* 1973. 156 p.

151. ROSENBERG, TERRY J. *Residence, Employment, and Mobility of Puerto Ricans in New York City.* 1974. 230 p.

152. MIKESELL, MARVIN W., editor. *Geographers Abroad: Essays on the Problems and Prospects of Research in Foreign Areas.* 1973. 296 p.

153. OSBORN, JAMES F. *Area, Development Policy, and the Middle City in Malaysia.* 1974. 291 p.

154. WACHT, WALTER F. *The Domestic Air Transportation Network of the United States.* 1974. 98 p.

155. BERRY, BRIAN J. L., *et al. Land Use, Urban Form and Environmental Quality.* 1974. 440 p.

156. MITCHELL, JAMES K. *Community Response to Coastal Erosion: Individual and Collective Adjustments to Hazard on the Atlantic Shore.* 1974. 209 p.

157. COOK, GILLIAN P. *Spatial Dynamics of Business Growth in the Witwatersrand.* 1975. 144 p.

158. STARR, JOHN T., JR. *The Evolution of Unit Train Operations in the United States: 1960–1969—A Decade of Experience.* 1976. 233 p.

159. PYLE, GERALD F. *et al. The Spatial Dynamics of Crime.* 1974. 221 p.

160. MEYER, JUDITH W. *Diffusion of an American Montessori Education.* 1975. 97 p.

161. SCHMID, JAMES A. *Urban Vegetation: A Review and Chicago Case Study.* 1975. 266 p.

162. LAMB, RICHARD F. *Metropolitan Impacts on Rural America.* 1975. 196 p.

163. FEDOR, THOMAS STANLEY. *Patterns of Urban Growth in the Russian Empire during the Nineteenth Century.* 1975. 245 p.

164. HARRIS, CHAUNCY D. *Guide to Geographical Bibliographies and Reference Works in Russian or on the Soviet Union.* 1975. 478 p.

165. JONES, DONALD W. *Migration and Urban Unemployment in Dualistic Economic Development.* 1975. 174 p.

166. BEDNARZ, ROBERT S. *The Effect of Air Pollution on Property Value in Chicago.* 1975. 111 p.

167. HANNEMANN, MANFRED. *The Diffusion of the Reformation in Southwestern Germany, 1518-1534.* 1975. 248 pp.

168. SUBLETT, MICHAEL D. *Farmers on the Road. Interfarm Migration and the Farming of Noncontiguous Lands in Three Midwestern Townships, 1939-1969.* 1975. 228 pp.

169. STETZER, DONALD FOSTER. *Special Districts in Cook County: Toward a Geography of Local Government.* 1975. 189 pp.

170. EARLE, CARVILLE V. *The Evolution of a Tidewater Settlement System: All Hallow's Parish, Maryland, 1650–1783.* 1975. 249 pp.

171. SPODEK, HOWARD. *Urban-Rural Integration in Regional Development: A Case Study of Saurashtra, India—1800–1960* . 1976. 156 pp.

172. COHEN, YEHOSHUA S. and BERRY, BRIAN J. L. *Spatial Components of Manufacturing Change.* 1975. 272 pp.

173. HAYES, CHARLES R. *The Dispersed City: The Case of Piedmont, North Carolina.* 1976. 169 pp.

174. CARGO, DOUGLAS B. *Solid Wastes: Factors Influencing Generation Rates.* 1977. 112 pp.

175. GILLARD, QUENTIN. *Incomes and Accessibility. Metropolitan Labor Force Participation, Commuting, and Income Differentials in the United States, 1960–1970.* 1977. 140 pp.

176. MORGAN, DAVID J. *Patterns of Population Distribution: A Residential Preference Model and Its Dynamic.* 1978. 216 pp.

177. STOKES, HOUSTON H.; JONES, DONALD W. and NEUBURGER, HUGH M. *Unemployment and Adjustment in the Labor Market: A Comparison between the Regional and National Responses.* 1975. 135 pp.

179. HARRIS, CHAUNCY D. *Bibliography of Geography. Part I. Introduction to General Aids.* 1976. 288 pp.

180. CARR, CLAUDIA J. *Pastoralism in Crisis. The Dasanetch and their Ethiopian Lands.* 1977. 339 pp.

181. GOODWIN, GARY C. *Cherokees in Transition: A Study of Changing Culture and Environment Prior to 1775.* 1977. 221 pp.

182. KNIGHT, DAVID B. *A Capital for Canada: Conflict and Compromise in the Nineteenth Century.* 1977. 359 pp.

183. HAIGH, MARTIN J. *The Evolution of Slopes on Artificial Landforms: Blaenavon, Gwent.* 1978. 311 pp.

184. FINK, L. DEE. *Listening to the Learner. An Exploratory Study of Personal Meaning in College Geography Courses.* 1977. 200 pp.

185. HELGREN, DAVID M. *Rivers of Diamonds: An Alluvial History of the Lower Vaal Basin.* 1979. 399 pp.

186. BUTZER, KARL W., editor. *Dimensions of Human Geography: Essays on Some Familiar and Neglected Themes.* 1978. 201 pp.

187. MITSUHASHI, SETSUKO. *Japanese Commodity Flows.* 1978. 185 pp.

188. CARIS, SUSAN L. *Community Attitudes toward Pollution.* 1978. 226 pp.

189. REES, PHILIP M. *Residential Patterns in American Cities, 1960.* 1979. 424 pp.

190. KANNE, EDWARD A. *Fresh Food for Nicosia.* 1979. 116 pp.

191. WIXMAN, RONALD. *Language Aspects of Ethnic Patterns and Processes in the North Caucasus.* 1980. 224 pp.

192. KIRCHNER, JOHN A. *Sugar and Seasonal Labor Migration: The Case of Tucumán, Argentina.* 1980. 158 pp.

193. HARRIS, CHAUNCY D. and FELLMANN, JEROME D. *International List of Geographical Serials, Third Edition, 1980.* 1980. 457 p.

194. HARRIS, CHAUNCY D. *Annotated World List of Selected Current Geographical Serials, Fourth, Edition. 1980.* 1980. 165 p.

195. LEUNG, CHI-KEUNG. *China: Railway Patterns and National Goals.* 1980. 235 p.

196. LEUNG, CHI-KEUNG and NORTON S. GINSBURG, eds. *China: Urbanization and National Development.* 1980. 280 p.